Snooperbase Waddington

SNOOPERBASE WADDINGTON

HOME OF THE UK'S INTELLIGENCE, SURVEILLANCE, TARGET ACQUISITION AND RECONNAISSANCE

MIKE HILEY

TEMPEST BOOKS

Published in Great Britain by Tempest Books
an imprint of Mortons Books Ltd.
Media Centre
Morton Way
Horncastle LN9 6JR
www.mortonsbooks.co.uk

Copyright © Tempest Books, 2024

All rights reserved. No part of this publication may be reproduced or transmitted in any form or by any means, electronic or mechanical including photocopying, recording, or any information storage retrieval system without prior permission in writing from the publisher.

ISBN 978-1-911704-21-8

The right of Michael Hiley to be identified as the author of this work has been asserted in accordance with the Copyright, Designs and Patents Act 1988.

Typeset by Jayne Clements (jayne@hinoki.co.uk), Hinoki Design and Typesetting

Contents

Introduction	7
Chapter 1 The History of RAF Waddington	8
Chapter 2 Nimrod AEW.3	15
Chapter 3 E-3D Sentry AEW Mk.1	22
Chapter 4 Operation Deliberate Force mission summary	52
Chapter 5 Nimrod R.1	56
Chapter 6 Sentinel R.1	68
Chapter 7 Shadow R	83
Chapter 8 RC-135W Rivet Joint	95
Chapter 9 ISTAR in a New Era: RPAS	115
Chapter 10 Royal Air Force Aerobatic Team: The Red Arrows	126
Chapter 11 Exercises and Deployments	146
Chapter 12 Air Traffic Control Recollections	181
Chapter 13 Visitors and Diversions	193
Chapter 14 Visiting ISTAR Assets	216
Chapter 15 Waddington International Airshow	221
Chapter 16 Preserved Aircraft	231
Chapter 17 Heritage Centre	238
Chapter 18 Lincolnshire and Nottinghamshire Air Ambulance Charitable Trust	242

Introduction

So, here it is. The finished article. It is amazing what can happen after a drink in the local one summer's evening in 2023 – I wonder if the low flying A400 that buzzed us had any influence on the route of conversation that night...

Living in Waddington, I knew I wanted to do something around the base. My initial thought was to concentrate on the ever-growing Cobra Warrior exercises that are hosted here, but realised the subject was somewhat limited and probably not enough to fill a book. Another chat in the local brought up a volume written by John F. Hamlin in 1996, For Faith and Freedom, depicting the history of RAF Waddington for the first 80 years and that was where the spark was ignited. Royal Air Force Waddington is now the UK hub for ISTAR – Intelligence, Surveillance, Target Acquisition and Reconnaissance and this has been the focus for the base since the mid-80s with the end of the V-bombers. It felt a like a good place to start, and so SNOOPERBASE WADDINGTON (thanks Graham!) was born, the name being a little nod to the Osprey Superbase series of books most aviation enthusiasts of the 1990s would have had in their collections.

Thanks to everyone who has contributed and helped me throughout, but an extra special thanks must go to two individuals: Ken Withers and Graham Robson, both of whom have produced some incredible images from their back catalogues to help bring the history of the station to life, particularly in the exercises and visiting aircraft sections. It is also worth noting that this book probably would not have happened without that chat in the pub with Graham, so thanks for the support – sorry for sending you down many rabbit holes looking for images!

I hope you enjoy reading this book as much as I have creating it. Maybe I will see you at the fence line one day?

Mike

The author in his (much) younger days, on Alpha taxiway at RAF Waddington, standing in front of 8/23 Squadron E-3D ZH107, on a typically wet English day in the early 90s, complete with binoculars and the traditional patch covered jacket.

CHAPTER 1

The History of RAF Waddington

RAF Waddington began its life as a Royal Flying Corps (RFC) training station in 1916, around two years before the formation of the Royal Air Force. A much smaller base than it is today, from north to south it ran the length of 'High Dyke', which is the road inside the base that links the two main entrance gates and from west to east the perimeter was along Mere Road. Mere Road is still there today, although public access heading east stops at the gate and then on the opposite side it stops at the A15.

RFC Waddington started with three hangars on the western edge and a further six on the south east corner with no set runway designations, aircraft just took off in whichever direction they needed to. The first RFC squadrons arrived in late 1916, with 47 and 48 Reserve Squadrons, flying MF.11 Shorthorn and Graham White aircraft respectively, which like many aircraft of the period were biplanes.

After the Great War, forward deployed squadrons had returned to RFC Waddington, although on their return little flying occurred and the station was put into Care and Maintenance and eventually closed for six years. In the early 1930s, the Government admitted that its reduction in military spending during the 1920s had not been the best plan and with the upsurge of Nazi Germany, it was announced in 1934 that the RAF would receive a major expansion, with several airfields to be rebuilt and made larger, Waddington being one of those sites chosen.

The airfield expanded east to the A15 and the five hangars that are still in use today were built, along with more support buildings, such as the Officers' Mess, NAAFI, Married Quarters and station offices, all of which still stand today. By 1939 Waddington was home to 44 and 50 Squadrons, as it would be throughout much of its history, flying Handley Page Hampden medium bombers.

The original NAAFI building in 2024. GRAEME ROBERTSON

On 3rd September 1939, the first day of the Second World War, Waddington-based aircraft carried out their first missions, with a dozen — six from each squadron, being sent to attack a fleet of German ships.

Lincolnshire was always a popular target for the Luftwaffe thanks to its multitude of bomber and training bases and during a raid in the early hours of 9th May 1941, the NAAFI (Navy Army Air Force Institutes) was one of the buildings struck by German bombs, along with the village church and the Horse and Jockey pub.

Sadly, a number of people on the base were killed, one of them being Mrs Constance Raven, who was the manager of the NAAFI at the time. The NAAFI was rebuilt and, in honour of Constance, named The Ravens Club. It still bears this name some 80 years later, although the club only occupies half of the building now, with the other half home to the RAF Waddington Heritage Centre. Remnants of dark green camouflage paint can still be seen on the outside of the building, as well as the different coloured bricks showing where the building was rebuilt.

A close-up of what remains of one of the 80-year-old bomber dispersals on the east of the base, taken in March 2024.

March 1943 saw the construction of the much-needed concrete runways, with the standard three runways built in a triangular pattern, along with adjoining taxiways and parking dispersals, some of which were to the east of the A15. Most of these dispersals have now gone, but some do still exist. One of them is the entrance to the Lincolnshire Fire and Rescue training centre, with the dispersals often used for periods of high visitors when the WAVE (Waddington Aircraft Viewing Enclosure) is full. The second is just to the north of the WAVE; spotters who came to Waddington pre-WAVE will no doubt remember the small pull-in where you could access the large pile of stones for a better look at aircraft arriving and departing.

The concrete runways came into use during November 1943 and with them came the arrival of Avro Lancasters at Waddington, with aircraft from 467 Squadron, who had been based at RAF Bottesford along with a new squadron – 463 Squadron, which was formed at

Waddington. Both squadrons were actually Royal Australian Air Force units and they remained at Waddington throughout the war, with their final mission of the war conducted on 25th April 1945, when Lancasters bombed an oil refinery in Norway.

The station switched to peacetime working hours before the end of May 1945, a much more relaxed affair than the previous six years, but this only lasted a little over two months. With victory in Europe achieved, there were still hostilities in the far east and to try and speed up an Allied victory the RAF Tiger Force was formed. This saw 463 and 467 Squadrons move out to RAF Skellingthorpe and Metheringham, respectively.

The special missions squadrons of Tiger Force were two of the RAF's most famous — 9 and 617 Squadrons, which arrived in June and July 1945. The two squadrons spent the next few months training before deploying east to India in January 1946. The next few years saw Avro Lincolns arrive at Waddington, with the type remaining at the base until 1953. Also present at various points during this period were B-29s, B-50s and C-47s of the United States Air Force along with a squadron of Washingtons (the RAF's name for the B-29) which were in Lincolnshire for ten months across 1951 and 1952.

The Jet Age had arrived by 1953 and RAF Waddington officially closed at the end of July for works to begin on extending and strengthening the runway, as well as the construction of the new Special Storage Area (SSA), also on the eastern side of the A15. The SSA is still in use today and forms part of the Lincolnshire Fire and Rescue Training Centre, with various training scenarios available across the site and in the buildings themselves. The work was completed by mid-1955 and Waddington now had a 9,000ft long main runway, with subsequent parking and lighting upgrades for

The sculpture outside Station Headquarters depicting a Lancaster and a Vulcan, which was created by James Sutton Sculpture for the station's centenary in 2016.

RAF Waddington as seen from the air in March 2023. On the right are the hangars, with the dark green Sentry (now Red Arrows) hangar viewable at the bottom corner, then the five hangars originally built in 1939 and lastly the Thompson building, which was built in the mid-90s. The wartime triangular pattern runways are still viewable, with the one in the bottom left now a parking location known as 'the lazy,' and the one halfway along the runway now a taxiway. The top of the image between the final two taxiways shows the extension made in the 1950s. In the very bottom left you can see the outline of six of the original 1939 bomber dispersals, as well as the WAVE car park, opened in 1996.

the expected new aircraft. The main runway had the headings of 21 and 03, with the other two runways from the 1943 upgrade no longer needed and becoming sections of the new parking locations.

The base also received a new control tower, which was one of only three built to the 7378a Pattern. Canberras of 21 and 27 Squadrons were the first jets based at RAF Waddington, although they were not the first jets to fly from there. The Avro facility at nearby Bracebridge Heath built two of the four Avro 707 test aircraft, with both aircraft having been towed down the A15 to Waddington for test flights before the base closed in 1953. The hangars, along with an angled tow ramp, are still viewable at Bracebridge Heath today.

The first Vulcan arrived at Waddington in August 1955 to carry out anti-radiation tests, although the first squadron was not active until mid-1956. Vulcans remained at Waddington throughout their service life, with the re-formed 44 and 50 Squadrons returning to the base, alongside 9 and 83 Squadrons and Number 230 Operational Conversion Unit, the latter conducting training for all Vulcan crews.

As the type was winding down in the early 80s, the Vulcan was called in to action during the Falklands conflict. Aircraft from the Waddington wing flew to Ascension Island to conduct the 'Black Buck' raids over the Falklands. Of course, the most famous of the Black Buck raids was the first one, with two Vulcans — XM598 and XM607 — and 15 Victor tankers in support, making an attack on the runway at Port Stanley. XM598 was the lead aircraft, but developed an issue, so XM607 completed the mission.

After the final operational Vulcans retired in 1984, Waddington became a little quieter, with only the Vulcan Display Flight remaining, plus a few years of the Joint Trials Unit (JTU) who were to test and trial the Nimrod AEW.3. As detailed later, the Nimrod AEW did not work out and the type was scrapped. This was the start of the main subject of this book, with what has now become known as ISTAR — Intelligence, Surveillance, Target Acquisition and Reconnaissance. Waddington then became home to Sentry AEW.1, Nimrod R.1, Sentinel R.1, Shadow R.1, and RC-135W Rivet Joint over the next few decades, with the most recent arrivals being the Protector RG.1 Drone and the Royal Air Force Aerobatic Team, the Red Arrows.

THE WADDINGTON AIRCRAFT VIEWING ENCLOSURE — WAVE
RAF Waddington has always been one of the busiest bases for visiting aircraft, so naturally it was a haven for spotters of all types, whether you were into photography, numbers or just watching, there was always something going on, more so in the late 80s and 90s with the various exercises. Parking was limited and people would park wherever they could, the main spot being a small pull-in just to the south of the runway on the A15. It could fit around half a dozen cars in and maybe a few more on the verge — until the police moved you on!

Another bonus to this location was the farmer's field just behind, which also had a few of the old dispersals in from the Lancaster days. One of these dispersals had a rather large pile of stones, which many spotters stood on for a better view. Waddington was getting busier and busier and it needed a dedicated space, which was compounded when someone was hit on the A15 and sent a fair few feet down the road by a passing car; fortunately the spotter survived. Thanks to some excellent and very persuasive work by a group of local spotters, who got the base as well as military companies such as Northrop Grumman to sponsor the site, the Waddington Aircraft Viewing Enclosure — the WAVE — opened on 29th March 1996.

A Belgian F-16 lands after completion of a training sortie in the late 90s. The photo is taken from the WAVE and shows the large pile of stones in the farmer's field, with some photographers needing even more height and placing ladders on top of an already well elevated viewpoint!

The WAVE had a cabin which contained a small shop with many kits, models, patches and books in one side and a café in the other. Information boards were placed around the car park that explained the history of the base and its residents and the place was very well received by the aviation enthusiast community. The original café/shop building was sadly condemned towards the end of 2022, with the current lease owner choosing to end the lease in the hope that some new blood would invest and bring something different. The WAVE closed at the end of January 2023.

Thankfully it was able to reopen in March under new ownership but with most of the same staff for the start of Exercise Cobra Warrior, with a small burger van which fed the hundreds of visitors during the busy exercise period.

CHAPTER 2

Nimrod AEW.3

RAF Waddington entered what was to be the ISTAR role with the arrival of the Nimrod AEW.3 (Airborne Early Warning) and the formation of the Nimrod Joint Trials Unit (JTU) in the mid-1980s, although the aircraft was already flying and already troubled by the time it arrived there. Waddington had been chosen as the main operating base for the Nimrod AEW as far back as 1979, with the British-built Nimrod being favoured over an American AEW aircraft, the E-3 Sentry, to replace the aging Shackleton AEW.2 aircraft operated by 8 Squadron.

One of the reasons Nimrod was chosen over Sentry, was that Nimrod could provide 360-degree visibility using one radar with two antennae, whereas the Sentry has a cone shaped blind spot under the aircraft that got wider the higher up the aircraft flew. The history of the RAF's and the UK's AEW assets had not been the smoothest, with old technology being used on current aircraft as a stopgap measure – the Shackleton AEW that Nimrod AEW was intended to replace was still using the same AN/APS 20 radar that was used on its predecessor, the Fleet Air Arm's Fairey Gannet. The AN/APS 20 dated back to the mid-1940s and was the outcome of a crash programme in the United Stated after the attack on Pearl Harbor in 1941.

The complexity of the Nimrod AEW project led to two companies acting as joint programme leaders – British Aerospace (BAe/BAe Systems) and GEC Marconi. British Aerospace were to deliver the aircraft and GEC Marconi the avionics and it was here that the issues began. Having two lead companies meant that whenever there was a delay in the programme, the default response was for each company to blame the other and it appeared at the time that the MOD did not really know what was going on.

British Aerospace was, however, able to deliver the aircraft on

Nimrod AEW XZ281 'DB3' (Development Batch) parked up at Waddington in the original white and grey scheme. GRAHAM ROBSON

time to allow flight characteristics testing, but the same could not be said about the avionics. The JTU formed from the Nimrod AEW training squadron, which was already active at Waddington and preparing lesson plans, and the Aircraft and Armament Experimental Establishment (A&AEE) Nimrod AEW trials team when the system issues did eventually come to light to be able to monitor company progress more closely.

The avionics were known as MSA – Mission System Avionics and they consisted of the radar, ESM (Electronic Support Measures), IFF (Identify Friend or Foe), navigation system and communications and datalinks that all fed in to a central data handling system. That information was then passed on to multi-function operator consoles for the nine-man mission crew to work with.

The initial ask was for the MSA to be able to track up to 400 individual targets, with fewer operatives than the E-3 Sentry, the plan being to rely on automation. But the hardware and software developed was not up to the job and the system struggled to cope with 100 tracks. The radar performance was not the best either; a poor antenna choice coupled with electromagnetic interference from the engines was corrupting the radar picture, with operators unsure if they were looking at a real world track or clutter. Despite the aircraft flying very well with the extra lumps at the front and rear of the aircraft, Nimrod was still the wrong platform to choose.

It was much shorter than the Boeing 707, which was the basis of

NIMROD AEW.3

An operator sits at the number 1 console on board Nimrod AEW. RAF WADDINGTON HERITAGE CENTRE COLLECTION

the E-3 Sentry, and had around half the maximum allowable weight of the E-3, yet it was expected to do the same job. There simply was not enough room for all the operators and their consoles, the radar, electronics, power generation equipment and cooling systems.

The Nimrod also had issues with its cooling systems. The design was intended to disperse the heat generated by the equipment through the aircraft's fuel systems using heat sink, but this needed sufficient fuel in the tanks to work effectively. Once half the fuel in the tanks had been used up, the mission systems had to be shut down. This severely hampered the time on station for Nimrod.

Another issue was the use of silicon oil as a coolant. This unfortunately leaked into the fuselage and impregnated the carpet. Silicon oil has a very low friction and so walking on any wet surface (such as the aircraft steps) was fraught with danger! You needed a firm grip on the handrails.

JTU initially had three aircraft to work with, and the spec of the three provided were around two months behind the company aircraft across at Woodford, but eventually all aircraft were present at Waddington.

The Government reopened the AEW competition in 1986 with the aim of trying to spur on the manufacturer, who had enjoyed a cost-plus profits contract and once again it came down to the E-3 Sentry and Nimrod. Despite a few tweaks to Nimrod, it became clear that the mission system required a complete redesign — so the Government went with the E-3 instead.

While all this was going on, the Shackleton ploughed on until Sentry came in to service in the early 1990s, more on which shortly. The Nimrod project was cancelled in December 1986 and the Joint Trials Unit was disbanded, with the aircraft dispersed to various locations and eventually scrapped. XV263 was flown to RAF

XZ285, the first production standard aircraft sits on the dispersal at RAF Waddington. GRAHAM ROBSON

Finningley and was used by the Air Engineer Squadron as a ground instruction airframe and was occasionally seen on taxy around the base. It appeared on static display at the base's airshows. XZ287 ended up at RAF Stafford, for use by TSW (Tactical Supply Wing) and was utilised during various exercises on the base. Oddly it was then partly buried, as seen in September 1995.

The AEW variant was instantly recognisable, with large balloon type bulges on the nose and tail of the aircraft, each of which housed an antenna — one looking forward and one looking backwards. XZ283 is seen here conducting an overshoot and parked on the dispersals at Waddington. GRAHAM ROBSON

AIRFRAME HISTORY

XV259 – P2. Cockpit saved, privately owned

XV261 – P8. Scrapped at RAF Lyneham, 1995

XV262 – P7. Scrapped at Abington, June 1992

XV263 – P3. Flew to RAF Finningley after programme cancellation and was used by the Air Engineer Squadron as a ground instruction airframe. Fuselage was used for fatigue tests at Warton/Brough as part of the Nimrod MRA.4 programme, Scrapped 2011

XZ280 – P5. Scrapped at Abingdon, April 1992

XZ281 – DB3. Scrapped at Abingdon, 1991

XZ282 – P6. Scrapped at Elgin, March 1996

XZ283 – P4. Scrapped Abingdon, 1991

XZ285 – P1. Scrapped Abingdon, May 1992

XZ286 – DB1. Scrapped Abingdon, 1991, rear section to Kinloss for ground instruction, Scrapped 1999

XZ287 – DB2. Scrapped at Stafford, 2014

P – Production Model

DB – Development Batch

The partly-buried XZ287. GRAHAM ROBSON

XZ287 parked on 'F' dispersal to the eastern side of Waddington's runway. Graham Robson

CHAPTER 3

E-3D Sentry AEW Mk.1

The ill-fated Nimrod AEW project was collapsing by 1986 and the RAF needed a suitable airborne early warning aircraft. There was only one obvious choice: Boeing's E-3 Sentry, which was already in service in the United States Air Force and the relatively new NATO AEW and Control Force.

The RAF originally ordered six E-3Ds in February 1987 and exercised an option for a seventh in October of the same year. One major difference between the RAF's new Sentries and the NATO and US versions already operational was the power plant. The RAF jets, as well as the French E-3Fs that were ordered around a similar time, were equipped with CFM56 turbofans, which were a much more fuel efficient and quieter engine than the Pratt & Whitney TF33s on the NATO and US jets.

The first E-3D Sentry AEW Mk.1 arrived at Waddington on 4th July 1990, but the first aircraft was not officially handed over to the RAF until 24th March 1991. The first flight in RAF hands occurred five days later and the aircraft officially entered service with 8 Squadron, on 1st July 1991, replacing the Avro Shackletons which had been around for 40 years and were a direct descendant of the Avro Lancaster. The Sentry is based on the highly successful Boeing 707 airliner, an aircraft which first flew in 1957 only eight years after the Shackleton's maiden flight and has seen military service in many guises. The most common types would be the KC-135 and KC-707 tankers used by many armed forces, but also the more specialist types such as the US Air Force and Royal Air Force RC-135, the E-6 Mercury of the US Navy and the very rare RE-3A of the Royal Saudi Air Force.

The arrival of the Sentry prompted the formation of the E-3D component of the NATO Airborne Early Warning (AEW) Force, with a new purpose-built hangar and dispersal for their use. The

8 Squadron Shackleton WL757 on Alpha dispersal, with ZH104 behind, on 1st July 1991. The aircraft provided a backdrop to the parade to mark the E-3D's entry into RAF service and thus began the start of the longest serving aircraft's history at RAF Waddington. PETER ROLT

dispersal was known as Alpha, with the hangar simply being the Alpha hangar. There was space for nine E-3 sized aircraft — a line of six bays on the northern side, two in the hangar with the final one just to the side, which also doubled as the wash pan.

The hangar provided enough height and specialist equipment to repair and maintain the signature 30ft diameter APY-2 radar, the performance of which has remained intentionally vague but which is able to see more than 250 miles at 30,000ft as well as having an air-to-ground mode should the need arise. The E-3D was crewed by 18 personnel — two pilots, navigator and flight engineer up front, and 14 mission crew in the main body of the aircraft.

ZH105 on static display in Alpha Hangar at the 2004 Waddington Airshow, showing the platform around the aircraft's radar as well as some of the emergency escape equipment that would be used should an aircraft have to ditch in the sea.

E-3D SENTRY AEW MK.1

Training of E-3D personnel was the responsibility of the E-3 training squadron, which was originally embedded within 8 Squadron. On 1st April 1996 it was renumbered 23 (reserve) Squadron. This was a change in pace for 23, having flown Lightnings, Phantoms and the Tornado F.3 previously. The reserve status was quickly dropped due to the front-line active service that the squadron became committed to.

23 Squadron markings were applied to the starboard side of the jets, with 8 Squadron markings being on the port side. On 1st September 2005, the recently disbanded 54 Squadron, who used to fly Jaguars out of RAF Coltishall was stood up once again and became the ISTAR OCU – Operational Conversion Unit – and training of all ISTAR aircrew then fell to them. 54 Squadron crests started to appear on the starboard side of the fuselage, just under the cockpit. 23 Squadron was disbanded on 2nd October 2009, the fleet then beginning to slowly lose their 23 Squadron markings, with only ZH105 still retaining fighter bars on the fuselage, although very faded, right to the end.

8 Squadron fighter bars.

54 Squadron and NATO AEW Force crest.

The starboard side of ZH103, showing the red/blue fighter bars and bright red eagle from the 23 Squadron crest.

ZH104 in 80th Anniversary markings lining up on Runway 21 as it was back then, in summer 1995. During the Sentry's service, several squadrons' anniversaries occurred with aircraft receiving special schemes. The first, as seen here, was in 1995 when ZH104 was in a simple scheme to celebrate 80 years of 8 Squadron. It had enlarged full colour fighter bars and a large Arabian dagger, known as a Jambiya, from the Squadron's crest. The Jambiya was adopted in recognition of the unit's association with Arabia in its early years. GRAHAM ROBSON

Ten years passed before the next special scheme. This time ZH103 was the chosen frame. As both Sentry squadrons were active at this time and both being 90 years old, similar schemes were applied to each side of the aircraft. The starboard side (above) was in the colours of 23 Squadron and port side (below) had 8 Squadron, with a rippled flag effect in the squadrons' colours applied to the tail fin. Top image: GRAHAM ROBSON

The third special scheme was applied to ZH106, for the 100th anniversary of 8 Squadron in 2015. The scheme was again applied to the tail and consisted of the Jambiya in front of a large figure 8 in the squadron's colours, depicted as smoke trails from aircraft the squadron had flown in its history.

ZH106 was rolled out of Alpha hangar at Waddington in its new scheme on 7th April 2015 during works to remove the 'hump' in Waddington's runway. At this time the E-3D fleet was operating out of RAF Coningsby, a few miles to the east of Waddington. 106 made the short flight to Coningsby on the evening of 9th June and the scheme lasted until March 2016 when the jet departed to Manchester to be stripped ready for maintenance.

ZH106 seen at RAF Coningsby on the evening of 9th June 2015 in glorious summer evening light. GJ PLATT/BCAR IMAGES

The final scheme was ZH107, which was painted up to mark the 100th anniversary of 54 Squadron. However, the aircraft never flew in this scheme. As there was no Waddington airshow after 2014, images of it anywhere other than parked on Alpha, taken from the roadside, are almost non-existent – but the scheme is covered later in the book.

ZH106 returned from Manchester on 17th April 2016 and spent the next 11 months at Waddington undergoing the required maintenance in Alpha hangar. Once complete, 106 departed back to Manchester for repaint. The jet is seen here leaving Waddington in a very unusual colour for repaint on the afternoon of 24th March 2017.

The Sentry was often deployed in support of RAF and NATO operations as well as exercises throughout the world, such as the massive Red Flag at Nellis Air Force Base, flying in the vast Nevada Test and Training Range north of Las Vegas. The aircraft flew operationally in the Middle East, the Caribbean, Libya and in support Operation FORTIS, which was the first operational deployment of the HMS QUEEN ELIZABETH Carrier Strike Group in 2021. When operational tempo allowed, the E-3D was part of the rather spectacular RAF Role Demo display, seen at airshows up and down the country. ZH107 is seen here with two Tornado F.3s to the front and two Tornado GR.4s to the rear at RAF Waddington's airshow in 2008. The Role Demo was known as Exercise Summer Lightning and was a scenario based on the RAF attempting to take control of a disused air field. Also involved were two Hawk T.1s, a Chinook and a C-130 Hercules as well as some rather large pyrotechnics!

ZH104 on static display at the 2009 Waddington airshow; clearly visible is the Lincoln coat of arms under the cockpit. The heraldic device dates back to 1959 when RAF Waddington was granted the freedom of the City of Lincoln and, as a result, the base's aircraft were allowed to wear the City's coat of arms. The shield has been seen on Sentinel, Sentry and RC-135 aircraft, and of course the mighty Avro Vulcan, which called Waddington home from 1957 until 1984. The much-loved XH558 remained active at Waddington until 23 March 1993 and still retains its crest at the time of writing, although it currently sits in the corner of the Doncaster Sheffield Airport with uncertain prospects for the future.

ZH105 on static display in 2010, minus all engines and with just the cowlings left — plus a large amount of yellow tape to seal off as much of the jet from the elements as possible.

The E-3D fleet was only ever upgraded once since entry in to service, between 1998 and 2000. It was decided in 2009 that the RAF fleet would not be upgraded to the block 40/45 standard and due to this the jets quickly became out of date and fell behind the much older US Air Force and NATO examples.

Costs needed to be cut and the number of active E-3s started to shrink by the end of the first decade of the new millennium, with the government announcing the fleet would be reduced from seven down to six. The chosen aircraft to be withdrawn was ZH105, simply because it was the next in line for a major service. Following its withdrawal, 105 began losing parts. It was seen at Waddington's Airshow in 2009 on static display, one engine already gone.

The following year, at the 2010 airshow, it had lost all its engines — with large blue circles filling the gaps where they once were and various panels and gaps covered in bright yellow tape. ZH105 had a large sign placed behind the crew door just aft of the cockpit, which read 'Ageing Aircraft Programme Laboratory' and was transferred to the Defence Science and Technology Laboratory (DSTL) and given the full title of Ageing Aircraft Programme Laboratory (Sentry).

It was used to support a wide range of Sentry-specific integrity tasks, such as studying hydraulic pipe corrosion, Auxiliary Power

From left to right, ZH105, ZH107, ZH101 plus two others. Note the blacked-out windows on ZH107 and lack of radar on ZH105.

Unit (APU) overheat damage and Environmental Conditioning System (ECS) duct recovery. Later that decade, ZH105 also became the Waddington Emergency Synthetic Trainer (WEST), while continuing to support its DSTL role. This was developed to reduce the need to use an active jet to conduct emergency drills. Obviously, the drills that had to be carried out while airborne could not be done with 105, but anything else could. They ranged from abandoning the aircraft quickly to fire, smoke and fumes drills – where crews were trained and tested in extinguishing anything when airborne. The old electronics were prone to overheating and with two compartments underneath the main cabin the drills involved a lot of climbing through small passages. Failures of hydraulic systems such as flaps or landing gear could also been trained and tested for on the WEST.

The next two to be withdrawn were ZH102 and ZH107. The withdrawal officially occurred in 2019, however, both had appeared withdrawn as far back as 2017. The first signs were blacked out windows and tape around certain panels and doors. ZH104 was next to go, departing to the USA on 15th January 2020 for storage. In February 2021, 104 was inspected by personnel from the US Naval

Ascot 7008, E-3D ZH101, touches down on runway 20 at RAF Waddington at 15:06, 4th August 2021, mission complete. MAX SHORTLEY

Air Systems Command team and subsequently sold to the US Navy for $15 million to become a trainer for their E-6B Mercury fleet. The E-6B mission fleet is always in high demand and having a dedicated trainer will drastically reduce stresses on the operational aircraft.

After the departure of 104 and the announced withdrawal of 102 and 107, the fleet was reduced to three aircraft: ZH101, ZH103 and ZH106. These airframes continued to fly, although struggled with unserviceability issues. The RAF announced plans to retire the surviving E-3D fleet on 29th July 2021 and it came quicker than most were expecting. The final operational sortie of the E-3D was completed on 30th July 2021 over Iraq as part of the counter-Daesh Operation Shader, bringing an end to almost 30 years of operational sorties. The two deployed jets returned in early August, with Ascot 7007 ZH103 returning from RAF Akrotiri on the 2nd, followed by Ascot 7008 ZH101 two days later.

The E-3D was officially retired on 28th September 2021, with a small parade at RAF Waddington and two rather spirited flypasts from one of the squadron's remaining jets. At that point it was

Upon arrival back at the parking bays, both ZH103 and ZH101 had a fire engine water cannon salute, something that has become tradition for a type's or pilot's final flight. 101 is seen here entering bays 1-9, operational tasking complete for the fleet. MAX SHORTLEY

unclear what the future would hold for the jets, but sightings of Chilean Air Force personnel in the local Tesco Express started some rumours about potential new owners. At the point of retirement, the three aircraft mentioned previously were still airworthy and active – ZH101, ZH103 and ZH106. They flew occasionally to keep the aircrew current, should the need for training a new owner be required, using the SOLEX callsign. The withdrawal of the usual NATO callsigns signified that the aircraft were no longer NATO assets.

The rumours were confirmed on 19th January 2022 with the Sentries to replace the Fuerza Aérea de Chile (FACh)'s sole AEW aircraft, the Phalcon, again an AEW aircraft based on the Boeing 707. It was initially thought that the three E-3s were being sold to Chile, with two fliers (103/106) and 101 as a spares package, but

Chilean Air Force Boeing 767-3Y0ER number 985 arrives at RAF Waddington on a cold and crisp January morning, under the callsign 'FACH 985' with personnel on board ready to commence E-3D conversion training.

it later came out that they were gifted; full details of the agreement are unknown. FACh crews arrived on board the sole FACh B767 early on 26th January 2022. Three months of ground school followed for the Chileans, with the first flight for them occurring on 4th May 2022 and the final flight just a little over two months later, on 12th July 2022. The FACh B767 returned a few days after to bring personnel over for the graduation events and then to take everyone home, the task of ferrying the aircraft left to the RAF crews. For reasons unknown, it also transpired that the transfer of ZH101 as a spares package did not materialise — it would remain behind.

The Chileans' first flight was in ZH106, callsign 'Solex 01' from Waddington on 4th May 2022. The sortie consisted of some brief general handling to the north of Waddington followed by numerous visual and radar circuits. The aircraft is seen here on one of many approaches to runway 20, with ZH101 and ZH102 (left) in the background.

SNOOPERBASE WADDINGTON

Throughout 2022, there were multiple flights in and out of RAF Waddington of FACh C-130H and KC-130H aircraft to collect spares and equipment to the delight of local spotters and photographers. This KC-130H version (above) arrived on 26th July 2022 using the callsign 'FACH 999'. Another of the visitors was KC-130R 992 (below — image: Les York). The final Chilean Hercules came through some three months after the E-3s departed Waddington, arriving on the 24th and departing on the 26th.

The planned departure day for the two E-3s was July 25th 2022, with ZH103 departing first and ZH106 30 minutes behind. However, when the day came only ZH103 departed, lifting off RAF Waddington's runway for the final time at 11:08, using the callsign 'Ascot 7001'. ZH106 followed the next day as 'Ascot 7006'. Word of both departures had spread via social media and the roads around Waddington were busy with people hoping to get one final glimpse of these aircraft which were the longest serving type at the Lincolnshire base.

Ascot 7001 ZH103 departing Waddington's runway 20 for the final time, as viewed from Runway 02 approach.

ZH103 arriving in Chile on the evening of 27th July 2022. SIMON BLAISE

FACh 905 lining up on Runway 17L at Santiago Arturo Merino Benitez International Airport, Chile on 19th September 2023. Clearly noticeable are the darker grey patches where RAF insignia and serial numbers have been covered. ANDRÉS ARANCIBIA

Once the RAF crews had delivered the aircraft to Chile, the Royal Air Force markings were removed. You can see on the image above a slightly darker shade of grey where the aircraft's serial and defining markings would be. Chile normally paints the rudders of all its aircraft dark blue, with a single white star, but in the case of the E-3 they have just darkened the rudder and applied the star. The dark blue rudder will be applied once the aircraft receive their first upgrade which will include a glass cockpit.

The delivery flights did not go all that smoothly. As mentioned earlier, ZH106 was delayed getting away but it also had issues en

FACh 906 departing Santiago Arturo Merino Benitez International Airport on 5th November 2022. GUSTAVO MARTINEZ

route and was stuck in the Azores due to an issue in the tail. The jet was stranded there for around a week while waiting for spares to be sent out. A nice place to be, but hugely frustrating for the crew.

With the two aircraft departing for Chile in July 2022, the end eventually came for the remaining four E-3D airframes, albeit many months beyond schedule, and they were broken up on the eastern side of the airfield in October 2023. The delay to the disposal was down to a couple of factors. When the scrapping first went out to tender, only one company applied and the MOD denied them the tender with no competition. It is understood that this company did not apply for it the second time round. The other factor was to do with missing parts that were listed on the tender, but no longer with the aircraft. It is unknown what happened to these items and whether they were returned or not by the time the final scrapping commenced in October.

All the scrapped airframes were parked at various parts of the airfield after the initial withdrawal in summer 2021, initially on bays 1-9 (old Alpha dispersal), then on bays 22-25 at the southern

ZH107 showing the 54 Squadron special tail scheme mentioned previously, seen on 10th February 2023 with tape and black plastic over the cockpit, suggesting the cockpit glass had been removed, the air-to-air refuelling probe above the cockpit is also missing. Various other panels and doors are also taped up but the aircraft is still relatively intact at this point.

end, before their final resting place of bays 14-17 in the middle of the airfield, to the east of the runway as can be seen on page 43. This photo was taken in March 2023, during Exercise Cobra Warrior 23-1. Also viewable are five Indian Air Force Mirage 2000s in the centre, six Finnish F/A-18C Hornets in the bottom right and a Saudi Air Force C-130H just above them. A temporary surface was laid to get the aircraft off the bays and on to the old disused runway, where the aircraft sat for some time. The E-3s' radars were first to be removed in late '21/early '22, then the rudder and tail fins by the middle of 2022, the final parts being the radomes, which were all removed in September 2023.

ZH107 was the very last B707 airframe off the Boeing production line. It was hoped that part of it could be saved – a cockpit section to a local or RAF museum for example, but it was the first to be scrapped on 4-5th October 2023. However, before it was fully scrapped various key parts were cut out, such as the fighter bars, serial numbers, and Lincoln Crest. These were to be repurposed as collectable 'tags' by American company JetEyes. Parts were also saved from ZH101 and ZH102. Another company that managed to get some of the saved parts was Timeless Aero, who

ZH107 seen on 3rd May 2023. Not much has changed externally since February, but work has started to remove panels from the tail.

The final image I got of ZH107 intact, taken 28th June 2023. The escape doors over the wings permanently open (or missing!) and the vertical stabiliser no longer attached. Work had also started on the bottom of the uprights that supported the trademark black and white radar dish — these apparently were saved from scrapping, potentially for 54 Squadron, but this has not been confirmed.

SNOOPERBASE WADDINGTON

ZH105 was second for the chop, with scrapping completed on 10th October 2023 — it appeared nothing was saved from this jet and nothing has appeared online, which backs this up, probably because 105 had not flown in at least 13 years. This image was taken around midday on 9th October.

ZH105 seen just under five hours later the same day, its cockpit well and truly smashed off and the aircraft cut in half just aft of the wings; a harrowing sound could be heard as they peeled back the fuselage skin.

Scrapping of ZH101 was next; the aircraft was pushed back on to the scrapping bay on the morning of 11th October, past the remains of ZH105 and ZH107, a smaller pile than you would think for such a large aircraft.

Work on 101 started two days later, with the tail section carrying over to the following week. A local general aviation flight on Saturday 15th October provided a view of the decimation from above. All that remained of ZH101 at this point was the tail plane and small section of the fuselage. You can also see a couple of sets of main undercarriage gear on the far left, a collection of tyres in the centre just to the right of the JCBs and various engine cowlings and two complete engines. ZH102 waits its turn on the right, still with engines but missing all cowlings. The scrapping of ZH101 was completed on Monday 17th October.

A sad sight. The remains of ZH107 and ZH105 as seen on 13th October.

were advertising their services to make these offcuts into tables, or something for the office or man cave.

There was then just over a week of little to no change on ZH102. The jet was last noted 'complete' at 11am on 25th October, however just a few hours later at 14:30, not much was left – they made short work of the final jet. The remains were scooped up and loaded into the trailers of lorries and driven out. It is not known what happened to those engines or the multiple engine cowling pieces. Maybe they will end up with the FACh or Armee de l'Air (French Air Force) who also operate the E-3? It would not be long before the sight and sound of an E-3 would return to Waddington; in fact it was before the end of the same month all jets were scrapped, with a French E-3F example arriving to participate in Exercise Atlantic Trident 23-1.

The E-3D fleet were known as the Seven Dwarfs, a tradition carried on from the Shackleton days of 8 Squadron, when all the aircraft were named after characters from The Magic Roundabout and The Herbs. Each of the E-3s was assigned a dwarf name, with a plaque and a picture of their namesake inside the crew area, near the rear entrance door. The jets were named as follows:

ZH101 – Doc
ZH102 – Dopey
ZH103 – Happy
ZH104 – Sleepy
ZH105 – Sneezy
ZH106 – Grumpy
ZH107 – Bashful

'Happy' as seen on ZH103 in 2002. GARY PARSONS

AIRFRAME HISTORY

ZH101 – Delivered 3rd November, 1990. Arrived as 'Able 01'. Last flew October 2021.
Moved on to scrapping pan 9am 11th October 2023. Fighter bars and various other panels cut out 12th October. Scrapped 13/16th Oct 2023.

ZH102 – Delivered 4th July 1990. Arrived as 'Boeing 162'.
Withdrawn in 2019. Scrapped 25th October, 2023. First to arrive, last to leave.

ZH103 – Delivered 10th January, 1991. Arrived as 'Mead 03'.
Departure to Chile, 25th July 2022, new serial 905.

ZH104 – Delivered 4th April, 1991. Arrived as 'Mead 05'.
Departed to the US, 15th January 2020, sold to the US Navy.

ZH105 – Delivered 7th June, 1991. Arrived as 'Mead 165'.
Became the Waddington Emergency Synthetic Trainer (WEST).
Withdrawn many years ago, believed last flew 2009. Scrapped 9th-10th October 2023.

ZH106 – Delivered 30th July, 1991. Arrived as 'Allen 06'.
Departure to Chile, 26th July 2022, new serial 906.

ZH107 – Delivered 23rd August, 1991. Arrived as 'Mead 107'.
Withdrawn in 2019. Last 707 airframe built, 1st E-3D to be Scrapped 4th/5th October 2023

Note: The delivered date is the arrival of the aircraft at RAF Waddington. Work was carried out by British Aerospace (Now BAe Systems) prior to official handover to the Royal Air Force.

ZH103 callsign Solex 02 making the final E-3D Sentry landing at RAF Waddington on the evening of 12th July 2022. The Chileans' training complete.

FACh 905 leads out a Chilean Air Force KC-135E and C-130H to participate in the Military Parade of Chile Flypast, which forms part of the celebrations for Chilean Independence Day, on 19th September 2022. The second E-3D is also viewable in the hanger on the right. GUSTAVO MARTINEV

CHAPTER 4

Operation Deliberate Force Mission Summary

By retired Wing Commander and former 8 Squadron E-3D Tactical Director, Chris Jobling

There was something different when I walked into the Operations Wing Briefing Room just before 4am that morning. A sense of urgency was clear among my fellow crew members, all bustling about refining the final details from the four hours of mission planning we had done the day before.

The plan: get airborne at 0600, top up with fuel from a USAF KC-135 tanker over Germany, relieve the nightshift, NATO E-3A sitting over Lake Balaton in Hungary, complete six hours of command and control on station and land in Aviano to start our two-week deployment; so, what was different? It would soon become clear.

We had flown hundreds of missions in the former Yugoslavia theatre of operations through Operation Maritime Monitor and Deny Flight and were very familiar with the concept of operations and could probably compete on a level playing field with native Bosnians in term of our knowledge of the geography of the country. "It's happening today, they are going in," said a member of my crew. Knowing what this meant I was initially overwhelmed by a sense of trepidation as I walked down the long narrow corridor to the classified briefing room in the operations headquarters building to learn more.

Over 1,000 missions by NATO air assets deploying live ordnance would take place during our on-station period. This was going to be a hell of a day but as I looked around the room at my 18 colleagues, I was completely confident in their ability to deliver, in spades. We had tons of experience in the theatre of ops and had all participated in high-end warfighting training exercises from Red Flag in Las Vegas through Cope Thunder in Alaska and Carrier Group workups in the Caribbean. We also had the best AWACS on the market. We were ready, 'combat ready'.

Final boxes ticked; mission documentation collected, including an enormous list of all the 1,000 aircraft that would participate that day. The Air Task Order (ATO) listed everything that would fly by type, callsign, mission profile, task, base, timings, refuelling bracket etc. If you weren't on the ATO then you didn't fly.

In addition, we had an intelligence update including the threats, not just to us but to all the participants. From a self-preservation perspective, it was the Serbian MiG-29s that sparked our interest and could certainly spoil our day. A short bus ride to Alpha Dispersal; documents by the boxful on board; pre-flight checks complete; the all-important rations for our gusting 10-hour day loaded and we were strapped in and ready to go.

Fully fuelled, fully serviceable and carrying no snags we listened intently on the intercom as the captain called for engine start — No. 1 engine was first, followed by No. 4, No. 2, and finally No. 3; all were running with a satisfying hum which allowed us to switch to internal power, produced by our eight generators and with a "Have a good trip gentlemen," the ground engineer unplugged and headed back to the crew room for a cuppa.

After a short taxi, we lined up on Runway 20. Cleared for take-off by air traffic control our four CFM56 engines powered us down Waddington's runway and at maximum all up weight of 335,000lbs, half of which was fuel, we rotated into the climb having used most of the available tarmac. Through 10,000ft in the climb pointing roughly in the direction of Hungary, we were cleared to unstrap. Time to get to work; it was to be a long day.

It takes about 40 minutes to configure all the complex systems for use with the initial load being taken by the three airborne technicians: radar, communications and displays. Until their configuration was complete, the rest of the mission crew were effectively deaf and blind. Dual mode, pulse, and pulse doppler radar, displays and communications suite serviceable — we had all the tools necessary to carry out our duties.

Still with 750 miles to go we couldn't yet see our area of responsibility on radar but using high frequency (HF) radio datalink we could receive and display the on-station E-3A's picture, enabling us to start building our situational awareness, but first we needed to refuel. Today, as briefed, the KC-135 out of Geilenkirchen also had an early start and was waiting for us on the pre-arranged refuelling track over Germany. As the only aircraft in the world to have the flexibility of refuelling from either the boom/receptacle or probe and drogue system, it could just as easily have been one of our VC-10s, but with a slower transfer rate we would have needed to plug in for longer.

Although a routine operation, air-to-air refuelling, particularly for big jets, always came with some excitement, even more so at night and in unfavourable weather. All crew strapped in, the interior lights dimmed and all non-essential equipment put into standby. However, with a very experienced flight deck crew it went without a hitch and we took on 40,000lbs of fuel, more than enough to complete the mission with plenty in reserve in case of an extension on station or diversion. The KC-135 crews loved us as a customer. Unlike with fast jet customers, they could get rid of large amounts of fuel to us very quickly; for them it made for a short day.

With 300 miles to go, we now switched from HF to the primary datalink—Joint Tactical Information Distribution System (JTIDS). This gave us greater clarity of the air and surface picture we were receiving from our sister Geilenkirchen-based E-3A, which was manning the M1 orbit until our arrival. It also enabled us to receive free text messages, such as the handover message: five pages of text in a standard format that gave us everything we needed to take over command and control responsibility for the Area of Responsibility (AOR) without speaking a word on any radio.

I agreed the on-station time with the Tactical Director on the other jet, which would of course be his off-station time. "Magic 84 on station, Bingo 16:30 and we are refuelling capable," was my announcement to the Combat Air Operations Centre (CAOC). At precisely 09:00 our numerous radios came alive with my fellow operators taking control of the various airborne, surface and ground-based assets.

A huge sigh of relief when we realised the other AWACS sharing the load positioned over the Adriatic was also a UK jet from Waddington. We spoke the same language and we thought the same so we could not have wished for better. We never did find out if it was purely by accident or because the opening salvo of Deliberate Force was to be such a challenging mission that the ATO was programmed to have the UK jets at the forefront. We preferred to think so because we were good and better than the US, French or NATO jets, in our view anyway.

The next eight hours were the busiest of my life. The sheer volume of traffic, the multiple simultaneous events across the AOR, the challenges of interpreting non-English speaking NATO combat assets all speaking more quickly due to some stress. Keeping the picture updated had never been more important as it was transmitted across the airwaves to multiple assets. A plethora of tankers on air-to-air refuelling tracks in the Adriatic had to be managed, deconflicted and serviced with receivers as fragged or sometimes bootlegged.

Already airborne and with their high-power jammers spoiling the adversary's day were the USMC Prowlers who were our next-door neighbours on the flightline in Aviano. They really were pivotal to the survival of NATO assets, operating feet dry (overland in the AOR) by jamming surface-to-air missile systems (SAMs) and disrupting the bad guys' air picture. Close Air Support (CAS) aircraft were being directed up safe lanes into hostile territory and chopped to the Forward Air Controllers.

We monitored their frequencies too and could hear some serious warfighting stuff: "Mavericks away", "Come south, your target south of the clump of trees, west of the brown field," said in an outrageous French accent. "What luck, have I hit my target." Serbian MiG-29s up from Banja Luca, always an attention getter. They were a threat to us and we took it seriously even though the Hungarian MiG-21s held QRA and would react to any border incursion. When they headed our way, we exercised our only line of defence, which was to turn and run away bravely for the Austrian border. We were known as a High Value Asset and were therefore a prime target. We knew an incursion was unlikely, but it would take just one maverick MiG-29 pilot out to make a name for himself to ruin our day. Send in the F-16s!

A French Mirage was down, struck by a SAM. No answer from

EDBRO 33. Were they safe? Anyone get their position? Do I have the capacity to coordinate a Combat Search and Rescue (CSAR)? It was another task we were best placed to conduct; with our crew of ten operators we did not have unlimited capacity and would have to shed something to make room—but this would be priority! "Weather is bad over the target area, I need to hold but need fuel, find me a tanker." Okay this is a bootleg so which tanker has spare fuel, is it boom or drogue and which track is he on? ESM detecting SA6, were they a threat to who, where? Nimrod R, also Waddington-based, calling real-time intel on adversary aircraft taxying. Combat helos and other intel and transport assets looking for threat warning to themselves; we had a God's eye view and they didn't.

Was our picture getting to the ground in CAOC Vicenza? Because of our Beyond Line of Sight (BLOS) range we relied on our sister ship in the Adriatic to relay our picture. Constant chat on SATCOM with the CAOC, effectively the boss, where all the difficult decisions were taken, although we did operate with considerable autonomy within the level of delegation outlined in Special Instructions (SPINS).

Most important was the Rules of Engagement (ROE). And in the background, what was the weather in Aviano? Is our replacement aircraft airborne yet? We needed to complete the five-page handover message; who on the crew has got time to do that?

Some eight hours later we called 'off station'. Magic 51 had the load; we were done for the day. I took off my headset and soaked up the silence save for the constant hum of the CFM56 engines which had now gone up a level as we accelerated back towards the Italian border, Aviano and a cold beer.

My ears ached from the head clamp I had worn for knocking on 9.5 hours. Paperwork all over the operator stations needed to be gathered up; I needed to write the misrep, describing what just happened… what did? I looked around at my fellow operators. As the adrenalin subsided the faces now showed weariness. Most enjoying the silence brought about by removal of the headsets. "NATO 54 (front end callsign) you are cleared own navigation to Aviano Airbase, descend FL 150," I heard through the headset speakers as it lay on my console. What a wonderful message.

Safely back on the ground in a rather warm Aviano Airbase our Sentries towered over the Italian airbase's numerous guests. USAF F-16s and F-15s, Canadian F-18s, USAF F-117s, and USMC Prowlers. All the Italian bases were full of deployed NATO assets not to mention the huge US Navy carrier positioned in the Adriatic that would disgorge its 100 or so aircraft daily to further add to the AWACS load.

We turned in the crypto and mission papers, attended a hot debrief and adjourned to the hotel in Pordenone for the real debrief over several cold beers. We would fly seven more missions like this one prior to returning home some two weeks later to RAF Waddington and normality. Aviano Airbase and Pordenone, where we were accommodated, became our second homes. By the end of hostilities involving the former Yugoslavia most of our aircrew were sporting 100 mission badges, some had 200 and even 300.

The 8 Squadron boss and his flight commanders in 1995. Left to right, Sqn Ldr Mike Clapham, Sqn Ldr Steve Gorton, Sqn Ldr Chris Jobling, Wg Cdr John Hall (OC 8 Sqn), Sqn Ldr Bob Wilkey, Sqn Ldr Nick Byatt.

CHAPTER 5

Nimrod R.1

Nimrod R came into being in the late 1960s, with three aircraft being ordered in 1969 off the main Nimrod MR (Maritime Reconnaissance) production line. The aircraft were first delivered to the RAF in 1971, minus any mission equipment due to the secrecy surrounding the platform. The fitting out of the Signals Intelligence (SIGINT) suite occurred at RAF Wyton, Cambridgeshire, which was to be the home of Nimrod R for over 20 years. The three original aircraft – XW664, XW665 and XW666 – were handed over to 51 Squadron as complete R.1 aircraft in 1974. The aircraft had a crew of 29: 25 mission crew in the rear of the aircraft, two pilots, navigator, and flight engineer.

Thanks to its excellent capabilities, Nimrod R.1 was often deployed around the world and was involved in every major conflict of the late 20th century and early 21st, with the final operational sorties conducted over Libya as part of Operation Ellamy only a few days before the type's withdrawal in June 2011. For most of its life, the aircraft was shrouded in secrecy and it was only really when the aircraft moved to RAF Waddington in April 1995 that more details began to emerge, even then they were vague and no photos of the inside of the aircraft were allowed until retirement.

For several years, the aircraft was described as a 'radar calibration' aircraft, when in truth it was designed to receive and record signals of interest that would be passed on to GCHQ (Government Communications Head Quarters). Should GCHQ be overrun due to a Soviet invasion, Nimrod R would assume the ground stations duties to enable the UK to continue to monitor the next move and try and be one step ahead.

Nimrod R.1, and with it 51 Squadron, arrived at RAF Waddington in April 1995 from RAF Wyton. The squadron had three jets, however 666 never made it to Waddington. It had been up at RAF

Nimrod R.1 was originally in the same hemp scheme of the maritime Nimrods when the aircraft arrived at RAF Waddington, but the fleet was painted in overall RAF Barley Grey with a light grey spine in the early 21st century. A 51 Squadron jet is seen on tow over to the ERP (Engine Running Platform) during Exercise Nomad in 2004.

Kinloss, near Inverness, for maintenance with the Nimrod Major Servicing Unit when the squadron moved. The aircraft caught fire during a test flight out of Kinloss on 16th May 1995, forcing the pilot to ditch the aircraft in the Moray Firth, where it broke in two and subsequently sank. Thankfully everyone on board survived thanks to the skill of the captain, Flt Lt Art Stacey, who for his actions was awarded the Air Force Cross. Most of the aircraft was salvaged and the cockpit section of 666 is now on display at the South Yorkshire Air Museum in Doncaster.

With the accident involving XW666 — affectionately known as

XW666 cockpit section on display at SYAM in May 2024. ANDY SHELTON

Damien in the RAF — 51 Squadron needed a new aircraft, and an existing MR.2 was earmarked for conversion: XV249. XV249 now has one less emergency exit for the rear crew, due to the mission systems in the back obscuring one of the overwing exits — the R.1 having fewer windows than an MR version due to the larger number of crew workstations.

XV249, as a recently converted but unpainted Nimrod R.1, taxies out for a test flight from Waddington in 1997. GRAHAM ROBSON

Where operational commitments, or lack of, allowed, Nimrod would participate in the opening station flypast for the Waddington International Airshow, alongside Sentry AEW.1 and eventually Sentinel R.1 once it was in service. XW665 is seen here on 30th June 2007 starting its take off roll, the four Rolls-Royce Spey engines spooling up with dark smoke billowing out of the rear.

XV249 had an overhaul at RAF Kinloss and was flown to BAe Woodford for stripping of all the anti-submarine warfare equipment by the end of 1995, and was delivered to 51 Squadron at Waddington in December 1996. The aircraft first flew as an R.1 on 2nd April 1997 and after a few weeks was declared operational, and the fleet was back to its original three aircraft.

The differences between Nimrod MR and Nimrod R were limited to the untrained eye – the main difference being the absence of Magnetic Anomaly Detector (MAD) boom on the back of an R.1.

Other differences included additional sensors on the R.1 in place of the massive searchlight on the edge of the starboard wing as well as a multitude of antennae on the top and bottom of the aircraft.

Another difference was the weapons bay. When parked with engines off, the weapons bay would always be open on an MR version due to the lack of hydraulic power, but with Nimrod R.1 the bay doors were always shut – the bay used for the bulk of the aircraft's detection systems.

As mentioned earlier, Nimrod R.1 was heavily involved in all the

XV249 flies along the crowd line at the 2010 Waddington Airshow, on 3rd July. You can clearly see the array of antennae both on top and below the aircraft. The two-half circle/dome shapes towards the outer edge of each wing however are not antennae — they both house a small wheel which would be used should the aircraft make a wheels-up landing, the black cover coming off when in contact with the ground to try and protect the wings of the aircraft. A question often asked at air shows was 'what antennae are these'?

major conflicts during its service life thanks to its excellent sensor suite and highly-skilled operators, and during my many visits to the base the sight of a 51 Squadron aircraft was a rare one. Areas of operation included Afghanistan, Iraq, the former Yugoslavia and, right at the end of its service life, Libya. Nimrod R.1 would also regularly frequent the Baltic Sea, just as its replacement, the RC-135, does today.

The aircraft's original planned out-of-service date was 31st May 2011, but it was given a 90-day extension due to the Libya conflict as there was not another RAF type at the time that could take over the Nimrod's sorties. XV249 returned home from RAF Akrotiri, Cyprus, on 23rd May 2011 and was replaced by XW664 which conducted the final month of operations for any RAF Nimrod type, with the MR.2 retired and the MRA.4 version scrapped the previous year in 2010. 249 remained active at Waddington, keeping crews 'current', flying local sorties on a number of occasions in June 2011, whilst

A rare sight (for me at least!) XW665 on taxy at Waddington, taken 4th July 2007.

A close-up of the goose just behind the cockpit.

A close-up of the port side of the tail, showing the Waddington station crest and the years the type operated from Waddington. Both images on this page were taken at Waddington Airshow 2011 where both 249 and 664 were on static display.

249 enters the runway at Waddington.

664 remined on operations for most of June.

To mark the type's retirement, Nimrod XV249 was painted in special markings towards the end of 2010. These were simple, but effective. On the port side of the tail was the RAF Waddington station crest, with the years 1995-2011 and on the starboard side was the RAF Wyton station crest, with 1974-1995 underneath. On both sides of the fuselage, just behind the cockpit a large red goose with the years of service of the type underneath.

I was present at Waddington on 8th June 2011 to witness the practice of the Queen's Birthday flypast, which was using Waddington as a substitute to fly over. Thankfully, XV249 also conducted a local flight, with multiple circuits, touch and goes and low approaches.

249 on final, thankfully the sun made an appearance!

Nimrod R.1 flight deck — a completely analogue cockpit which dates to the 1970s with the type's introduction. The captain would sit in the left-hand seat, with the co-pilot on the right. The third seat viewable would be for the flight engineer, who would monitor the performance of all four engines throughout the flight. Out of shot and behind is the navigator's desk.

Looking down the rear of the aircraft, showing the full length with the various operator seats.

My thanks go out to Phil and Steve Slater of East Midlands Aeropark who allowed me to visit their Nimrod R.1, XW664, to obtain these photos on a cold and foggy day in December 2023. Steve is ex-51 Squadron ground crew and was showing a lot of the internal systems working; he hoped to be able to get hydraulic power to the aircraft at some point in the future to allow visitors to move the flight control systems. I would highly recommend a visit if you are in the area.

AIRFRAME HISTORY

XW664 — Original R.1 build, replaced XV249 on operations over Libya for the final month in May 2011.
  Final RAF operational Nimrod of any type, withdrawn 28th June 2011. Final flight to East Midlands airport, 12th July 2011. Preserved at East Midlands Aeropark.

XW665 — Original R.1 build, final flight 27th October 2009.
  Scrapped in hangar at RAF Waddington, large section of nose and forward fuselage preserved in Sinsheim, Germany. Emergency exit door in a friend's garage, Branston, Lincoln.

XW666 — Original R.1 build. Ditched 16th May 1995 in the Moray Firth.
  Cockpit section preserved at South Yorkshire Air Museum, Doncaster.

XV249 — Built as an MR.1 Maritime patrol version, first flight 22nd Dec 1970.
  Upgraded to MR.2 during fleet wide upgrades. Converted to R.1 spec in 1997. Returned from operations over Libya 23rd May 2011, Withdrawn from service 28th June 2011. Final flight to Cotswold Airport 29th July 2011. Transport by road and preserved at the Royal Air Force Museum, Cosford.

XW664 and XV249 (rear) on static display for the final time at the RAF Waddington International Airshow, July 2011 just a few days after withdrawal from service.

CHAPTER 6

Sentinel R.1

The Sentinel R.1 story begins in the late 90s, when Raytheon was awarded a contract for an airborne stand-off system, otherwise known as ASTOR. The aircraft started out as a Bombardier Global Express business jet, that was fitted with mission systems supplied and maintained by Raytheon. Five of these aircraft were ordered, along with multiple mobile ground stations.

The first flight was conducted in May 2004 after two years of installation work by Raytheon, the second following in the middle of 2005. The aircraft became known as the Sentinel R.1 in RAF service, with 5 Squadron chosen as the unit to operate the aircraft. 5 Squadron had disbanded in 2002, having previously flown Tornado F.3s out of nearby RAF Coningsby. To start with, 5 Squadron only had a handful of personnel and they operated out of a small office at Waddington, with a laminated A4 sheet of paper on the door.

Sentinel ZJ692 on approach to Waddington on the afternoon of 5th April 2006 while on a test flight. The aircraft is unpainted, showing the bare metal of the fuselage and various adaptations to make the Sentinel platform. JIM BAKER

SENTINEL R.1

The first Sentinel R.1 arrived at RAF Waddington in 2007 to join 5 (Army Co-operation) Squadron, a few years after the first ground stations. The Army Co-operation suffix was added to identify the co-operation of the two forces, with the Army running the ground-based stations. 5 (AC) Squadron and their Sentinels were very busy, with aircraft often deployed throughout the world, the first deployment being made to Afghanistan in November 2008.

The Sentinel was also used in and around the Baltic states, monitoring military forces stationed in the area, something which is now carried out by the RC-135 aircraft of 51 Squadron. The public's first look at this new aircraft was at Waddington's airshow on the weekend of 30th June and 1st July 2007. ZJ690 is seen here landing on 30th June after participating in the traditional station flypast to open the show.

ZJ693 lining up to depart for a training mission on the morning of 7th July 2008, as seen from the on-base viewing area for the 2008 airshow departure day.

Raytheon was the prime contractor for the ASTOR programme, but other companies were subcontracted to supply various parts, such as Agusta Westland for the landing gear, L-3 communications for the ground stations and of course Rolls-Royce for the engines. The radar used on the Sentinel is an upgrade on the side looking radar used on the US U-2 aircraft — a very high-altitude surveillance and reconnaissance aircraft, often seen above Waddington as it flies to and from its area of operation. The radar could produce photographic quality images which could be transmitted in real time to the ground stations via secure data links. The aircraft had a crew of five — two pilots and three mission crew and was capable of a mission endurance of more than 14 hours thanks to its fuel-efficient Rolls-Royce engines and large wingspan.

All the Sentinel aircraft originally had squadron markings to

The aircraft is seen here parked at the southern end of the airbase during the 2013 Waddington Airshow. KEN WITHERS

give them some identity, which consisted of a red bar across the top of the tail fin, with a Maple leaf in the middle. The Maple leaf commemorating the squadron's close links with the Canadian Corps during the First World War. During 2013, ZJ692 received a special scheme to celebrate the squadron's centenary — a bright red tail and large green Maple leaf. ZJ692 opened the airshow in 2013, conducting a flypast with the Royal Air Force Aerobatic Team — the Red Arrows.

As the operational tempo for 5 Squadron and their Sentinels increased, the aircraft slowly began to lose their squadron markings and matt paint from around 2013, and four of the five aircraft were painted in a darker, more glossy shade of grey devoid of any unit identification — a sign of the times for much of the RAF's aircraft.

Despite the capability and performance of the aircraft, it was announced in the 2010 Strategic Defence and Security Review that the Sentinel would be retired in 2015, as it would no longer

ZJ694 is seen lining up for departure on the morning of 11th September 2019 to participate in a Cobra Warrior sortie. Although the paint scheme was very low key, the jet still looked smart.

be required to support operations in Afghanistan with the withdrawal of troops expected in 2014. This decision was reversed in 2014 by the then Prime Minister, David Cameron, with the Sentinel expected to remain operational and in service until 2021. However, it was announced one aircraft was to be withdrawn with effect from 1st April 2017.

ZJ693 is seen here departing Waddington two days before that announcement on 29th March, its destination Hawarden airport, where it spent the rest of its RAF life. ZJ693 was the chosen aircraft, presumably as it was next in line for a service and repaint — saving the UK money since that work no longer needed to be carried out. At this point, 693 was the only aircraft still in the original 5 Squadron markings, and as can be seen here the aircraft was not exactly looking in pristine condition.

ZJ690 on 4th July 2007.

ZJ691 performing a flypast with the Red Arrows at the RAF Scampton Airshow, 10th September 2017. Scampton was an attempt to bring a full military show back to Lincolnshire, after the Waddington show was cancelled following the 2014 event. Although it was run by the same team who organise the Royal International Air Tattoo at RAF Fairford, it was not successful.

The Sentinel was continuously deployed for 12 out of its 14 years in service. It was involved in Operation Ellamy in Libya in 2011, where the aircraft's longest flight of 12 hours and 30 minutes was achieved, over Mali in 2013 and of course Afghanistan for several years. Possibly the most well-known and highly publicised mission of the Sentinel was in 2014, when the squadron deployed to Ghana as part of Operation Turus to assist in the search for the schoolgirls who had been kidnapped in Nigeria.

With the threat of retirement always looming overhead, the Sentinel was not upgraded during its service life and was quickly becoming obsolete. The Sentinels were retired from RAF service in March 2021, with the final operational sortie being carried out by ZJ694 on 25th February 2021, 5 Squadron was then disbanded at the end the March.

ZJ692 departing RAF Waddington on the morning of 4th September 2019.

Within a few months, the aircraft were stripped of any operational equipment and parked at the far corners of the airfield, tied down with large concrete blocks to prevent them tipping or blowing over due to the lack of weight. ZJ691 is seen here on 28th August 2021, with silver tape covering many of its seals — its fate then unknown.

SNOOPERBASE WADDINGTON

The first aircraft departed on 7th June 2022, with the last being N690BD on 21st March 2023. N691BD is seen here departing on 13th July 2022 bound for RAF Fairford to be on static display at that year's Royal International Air Tattoo.

It was announced in November 2021 that the aircraft were to be sold to the US Army. They were moved to the civil US register in the first half of 2022 and registered to Springfield Air – a private company specialising in the charter and sales of business jets. As each jet got its new serial applied, it also lost any of its remaining RAF insignia. With the end customer reported as being the US Army, it is expected that the aircraft will end up becoming E-11A BACN (Battlefield Airborne Communications Node) or similar platforms, but at the time of writing nothing has been confirmed.

Each aircraft had at least one test flight before departing to the US. The flights where basically to check the functionality of all systems prior to the long flight west. Each flight started with a high-speed taxi down the runway to check the brakes and reverse thrusters, if all was clear the aircraft would taxi back round to the threshold for take-off. The flights were relatively short, N694BD is seen here turning off Delta taxiway after a successful test flight with the RAT deployed just under the cockpit. The RAT (Ram Air Turbine) is normally only deployed in emergencies during a sudden loss of power. The RAT spins in the airflow and is connected to a hydraulic pump or electrical generator to provide power to essential flight controls. The reverse thrusters are also deployed on the rear of the engines.

As mentioned earlier, ZJ693 was the only aircraft to remain in its original scheme. It's seen here marked up as N693BD in August 2022 still with its red bar, but the 5 Squadron maple leaf along with other RAF insignia has been removed.

Sentinels in Arizona. N694BD, with N693BD directly behind, and just off to the right is N691BD, taken during February 2023 at Tucson, Arizona. BOB FERNADES

AIRFRAME HISTORY

ZJ690 – Arrived 24th January 2007, as 'Ascot 4774'.
Registered as N690BD and departed to Hawarden airport, UK 21st March 2023.

ZJ691 – Arrived 19th March 2008, as 'Ascot 9554'.
Registered as N691BD and departed to participate in the Royal International Air Tattoo at RAF Fairford on 13th July 2022. Departed to the US 18th July 2022.

ZJ692 – Arrived 21st October 2008, as 'Ascot 5132'.
Registered as N692BD and departed to the US 7th June 2022.

ZJ693 – Arrived 26th September 2007, as 'Snapshot 1'.
Departed to Hawarden Airport 23rd May 2017, returned to Waddington 25th February 2021 still with its 5 Squadron bar on the tail. Registered as N693BD and departed to the US 27th September 2022.

ZJ694 – Arrived 7th November 2008, as 'Sentinel 01'.
Registered as N694BD and departed to the US 20th July 2022.

N690BD parked on Bay 1 awaiting an engine run on in early 2023. The last aircraft to depart.

CHAPTER 7

Shadow R

The introduction of the Shadow R.1 into RAF service occurred in 2009, after an Urgent Operational Capability (UOC) requirement to support intelligence gathering in Afghanistan. Originally there were four aircraft and rather than a new squadron being formed for them, they operated as a flight within 5 (AC) Squadron at Waddington, simply known as Shadow flight.

The Shadow is based on the Beechcraft King Air 350, which is one of the many variants of King Air that Beechcraft makes. The original King Air aircraft first appeared in the late 1960s, with the King Air 200 arriving some 20 years later, which was the first 'Super' King Air, although the Super designation was dropped in 1996. Four aircraft were delivered between May and December 2009, after conversion from standard King Air to Shadow spec at Raytheon's facilities at Hawarden Airport, Flintshire.

These original four aircraft were numbered ZZ416, ZZ417, ZZ418 and ZZ419. Due to the nature of the aircraft and its role, very little is known about it. A Shadow has, at the time of writing, never appeared on static display at an airshow or RAF Families Day, but they have participated in flypasts over London to commemorate Royal events. Military personnel cannot get anywhere near it unless directly involved. When the type was to be introduced, the Commanding Officer of Base Operations was approached to ask if Shadow could be housed at Waddington. He was given very little detail on it initially, with no idea as to what it would be doing. Eventually he did get some details, which ended up with Shadow being housed at Waddington with the majority of the ISTAR assets.

Shadow is a twin-engine turboprop aircraft, with excellent speed and endurance and can operate up to 35,000ft. It is equipped with high-definition electro-optic and electronic sensors, enabling its crew to gather data and analyse it while airborne, or transmit via satellite to other assets or ground stations for analysis. The Shadow is not equipped with any weapons, but it does have a defensive aids system which consists, among other things, of sensors on the nose and tail of the aircraft to detect incoming missiles and it can also carry flares to divert the attention of heat-seeking missiles.

Shadow R.1 ZZ416 taxies out on 8th June 2011, still looking fresh and new. The missile detection sensors can be seen on the nose and just forward of the aircraft's serial number towards the rear.

During 2011, the Shadow flight of 5 Squadron took on a new number plate: that of the recently disbanded 14 Squadron, another ex-Tornado unit. With this came the 14 Squadron callsigns, such as Snake and Scarab. The fleet also grew to five aircraft in 2011, with the addition of ZZ504. A sixth aircraft arrived in 2013, in the form King Air 350 G-LBSB. This was a plain white aircraft, with no military markings at all, and it acted as a crew trainer for five years. All aircraft have a lifespan of landings, including touch and goes, before needing a rework and servicing, so having a non-mission capable aircraft for training sorties reduced the strain on the operational fleet.

King Air 350 G-LBSB operating at Waddington on 15th March 2017.

During the Strategic Defence and Security Review of 2015, Shadow was moved from the UOC requirement status it had been in since 2009 into the main core of the Ministry of Defence, along with the announcement of a sixth aircraft to add to the fleet. This was G-LBSB and it was sent for conversion in 2018. It was also announced that all aircraft would be upgraded to Shadow R.1+ spec, with additional sensors and capabilities added. The first aircraft to receive this new upgrade was ZZ417, returning to Waddington in 2019. Externally, the only real way to distinguish an R.1+ from a standard R.1 is the small budge at the top of the tail and an additional antenna aerial just above and to the left of the aircraft's serial number, as can be seen here on 23rd August 2019.

SHADOW R

R.1+ ZZ507 lines up on runway 02, 25th August 2021.

During 2017, a further two civil King Air 350s joined the fleet for training purposes, with one of these appearing on static display at the Royal International Air Tattoo 2018. There was even a sign in front giving the spec of a Shadow – the closest a Shadow has got to an airshow appearance!

Another 'upgrade' applied to the fleet between 2019 and 2022 was a new 'executive' paint scheme – a gloss white with light grey undersides and a red and dark blue cheat line from nose to tail, with a very cleverly split out RAF roundel and fin flash incorporated into it. To the uninitiated, it would look like every other business aircraft, which is exactly what the MOD wanted – to hide in plain sight. Not all aircraft received the paint scheme upon upgrade, with some returning to Raytheon for repaint once the front-line fleet had enough operational aircraft to meet demands. The final aircraft to upgraded to R.1+ was ZZ504, which was redelivered in August 2022.

It was always rumoured that the two additional civil King Airs would end up being converted to full Shadow specification, with the serials known as far back as 2019, and this was finally confirmed in November 2021 with the Shadow fleet increasing to eight. It was also announced that all aircraft would eventually be upgraded to R.2 specification. In the interim before the first fully fledged R.2

R.1+ ZZ416 lines up for runway 20, 6th October 2022.

aircraft are in service, a minor upgrade will be carried out to some, with these aircraft receiving the R.1++ designation. ZZ418 was the first, and at the time of writing, only aircraft to have received this upgrade, arriving at Waddington eight months after the final R.1+ upgrade was completed. Externally, there is no difference between the R.1+ and R.1++, but it is believed that the upgrade carried out will make it far quicker to convert the aircraft to full R.2 specification. As it stands, the first R.2 model is expected sometime in 2025.

Shadow has been used operationally on an almost continual basis since its introduction, both in the UK and overseas, obviously starting with Operation Herrick in Afghanistan, which was the reason the UOC requirement was initially raised. A milestone 10,000 hours of operational flying was surpassed in 2012, showing how important and in demand this asset was. Back in the UK, Shadow is very active, with multiple sorties most days. Little to nothing is known about these sorties; they could be operational or training, but what is known is that they are often seen via internet tracking platforms orbiting major towns and cities, sometimes for hours on end.

One widely publicised group of missions occurred in 2020, with Shadow operating over the English Channel to observe and track incoming migrants who would often put themselves in grave danger, trying to cross the busiest shipping lane in the world in small overloaded boats with no radios. Surprisingly the high-tech

R.1++ ZZ418 on final to Waddington on its delivery flight on 19th April 2023, flown by a Raytheon crew, using the aircraft's number as its callsign — 'ZuluZulu418'. Viewable are the multitude of sensors and optics under the main body of the aircraft.

Shadow with its massive suite of sensors was not the first choice for the British Government. The large A400 transport aircraft was first selected to operate over the channel, presumably counting on the pilots' and loadmasters' mark one eyeballs to track the migrants!

With the retirement in mid-2021 of the Army's fleet of Islanders, which had held a constant presence in Northern Ireland for

ZZ504, the last remaining grey R.1 spec lines up on Runway 20 at Waddington on 26th October 2020.

many years, Shadow became a familiar sight to those around the military base of Joint Helicopter Command Flying Station (JHC FS) Aldergrove, formally RAF Aldergrove, with the aircraft being stationed there for weeks at a time. Eventually, Shadow visits to Aldergrove became less frequent and when visits were required, they tended to operate out of Waddington and back, usually using the 'Minx' tactical callsign.

In the last quarter of 2023, Shadow was once again deployed

Shadow R.1+ ZZ419 enters the runway on 22nd September 2022, using the callsign 'Vulcan 49'. Vulcan is the station callsign and any based aircraft can use this. Since introduced, Shadows have always used part of their serial as the callsign, which helps identify which aircraft is airborne at the time — an unusual but helpful occurrence. The only variations to this are when the aircraft deploy overseas and they use an Ascot or RAFAIR callsign, or the sortie is run by 56 Squadron with their Firebird callsign. 56 Squadron is the ISTAR test and evaluation squadron.

to RAF Akrotiri and it was believed the deployment was related to the ongoing issues and tensions in Lebanon, Israel and Gaza. Two aircraft were deployed, ZZ507 and the top spec ZZ418 on 25th October, routing via Naples, on the western coast of Italy. Operational sorties began within a week, with the aircraft flying orbits over the eastern Mediterranean and off the Lebanese coast. It was then announced in early December 2023 that Shadow would be assisting in locating the remaining British hostages that Hamas had taken two months earlier.

AIRFRAME HISTORY

ZZ416 – Ex G-JENC, delivered to Waddington 27th May 2009 as an R.1.
Departed 10th June 2021 for upgrade. Delivered to Waddington as an R.1+ 9th February 2022.

ZZ417 – Ex G-NICY, delivered to Waddington 7th July 2009 as an R.1.
Departed 31st October 2018 for upgrade. Delivered to Waddington as an R.1+ 17th June 2019. Departed 8th March 2023, presumably for upgrade, unsure if R.1++ or R.2 spec.

ZZ418 – Ex G-JIMG, delivered to Waddington 3rd September 2009 as an R.1.
Departed 9th October 2020 for upgrade. Delivered to Waddington as an R.1+ 10th June 2021. Departed 13th October 2022 for upgrade. Delivered to Waddington 19th April 2023 as the first R.1++.

ZZ419 – Ex G-OTCS, delivered to Waddington 10th December 2009 as an R.1.
Departed March 2018? for upgrade. Delivered to Waddington as an R.1+ 4th October 2018.

ZZ504 – Ex G-CGUM, delivered to Waddington 9th December 2011 as an R.1.
Departed 26th August 2021 for upgrade. Delivered to Waddington 1st August 2022 as an R.1+.

ZZ505 – Ex G-DAYP, arrived at Waddington as a dark green King Air 350 29th September 2017.
Departed 11th June 2019 for conversion. Planned be delivered as a Shadow R.2.

ZZ506 – Ex G-GMAD, arrived at Waddington as a dark grey/white King Air 350 26th September 2017.
Departed to Bournemouth Airport on 15th April 2020. Departed to Hawarden 7th April 2022 for conversion. Planned be delivered as a Shadow R.2.

ZZ507 – Ex N5055U, delivered to Waddington as G-LBSB 30th July 2013 as a white King Air 350.
Departed for conversion to Shadow in March 2018. Delivered to Waddington 4th November 2019 as an R.1+.

Shadow R.1+ ZZ504 on finals on a perfect summer's day in August 2022.

SNOOPERBASE WADDINGTON

Shadow R.1+ ZZ507 taxies down RAF Waddington's Delta taxiway on a cold February morning in 2023. Shadows hardly ever use Delta taxiway and in all the years they have been stationed at Waddington this is the only time I have ever photographed one coming down Delta. The shot was taken on 7th February 2023. The aircraft was conducting flare trials at RAF Donna Nook, which is a bombing range on the east coast of England, just south of Cleethorpes. If an aircraft has flares on board, it is considered an armed aircraft and extra safety measures are put in place to protect infrastructure and personnel, with the aircraft parking as far away from the main base as possible. In Waddington's case that comes down to bay 18 Alpha, which is the centre of a large hardstanding in the south eastern corner of the airfield.

CHAPTER 8

RC-135W Rivet Joint

The UK Ministry of Defence announced in March 2010 that an agreement had been reached with the United States to purchase three RC-135W Rivet Joint airframes and that they were to be known in Royal Air Force service as Airseeker R.1. The announcement came around 18 months after the United States Defense Security Cooperation Agency had originally notified US Congress of a possible sale to the United Kingdom of three former United States Air Force KC-135R Stratotankers that would be converted into RC-135W Rivet Joint aircraft.

The aircraft were to replace the ageing Nimrod R.1 aircraft, which were originally intended to be replaced by a SIGNIT version of the Nimrod MRA.4, which in 2008 was already millions of pounds over budget and seven years late. At this point the RAF had started to lose interest, which is why another platform was required, and there was only one viable option. The deal was to kit out and convert the three KC-135R aircraft to a setup almost identical to that of the USAF RC-135W fleet, a fleet with which RAF personnel were already familiar, having had aircrew on three-year exchange postings to Offutt Air Force Base with the 55th Wing since the early 2000s. Since the RAF crews were already familiar with the Rivet Joint term, the Airseeker name didn't really take off (no pun intended!). It faded away and appears to be completely dropped.

The purchase deal included, among other things, tools and test equipment, personnel training, US government and contractor support and a mission trainer. Once the deal had been confirmed, and the Nimrod R.1 retired in 2011, the RAF started to send additional aircrew and groundcrew to Offutt for training in preparation for them to work alongside their US counterparts until the RAF aircraft were ready.

The Rivet Joint conversions were done by L3 Communications,

One of the first visits of an RC-135W Rivet Joint after the announcement that the UK would buy three examples, was in July 2010 for that year's airshow, with 62-4130 'OF' seen here arriving for static display on 1st July 2010.

based in Greenville, Texas with work on the first aircraft, which was to be ZZ664, starting in early 2011. With all the aircraft previously being air-to-air refuelling tankers, the first job was to remove all the equipment used for this role, such as the boom at the rear. The aircraft was then stripped back to bare metal to enable L3 engineers to begin the lengthy task of fitting the many different sensors and aerials, including its signature elongated nose. By summer 2013, ZZ664 emerged in the traditional 'white top' colour scheme of the Rivet Joint force, with 'ROYAL AIR FORCE' on the side in the same block font type of the US jets.

ZZ664 arrived at RAF Waddington on 12th November 2013 and it was flown direct from Greenville by L3 aircrew. On final approach to Waddington, 51 Squadron aircrew took over control and landed the aircraft in front of assembled personnel and VIPs. The aircraft was hangared not long after arrival, as specialists from Qinetic at Boscombe Down and the UK MOD needed to certify that 664 was fit for operations and above all, safe to fly. This was thankfully achieved and ZZ664 returned to the skies in May 2014, with

ZZ664 taxies out from bays 1-9 on 3rd July. KEN WITHERS

A close-up of the nose and cockpit section of ZZ665 when the aircraft was on static at the Royal International Air Tattoo as part of the RAF100 display in July 2018. To date, this is the only occasion when an RAF RC-135W has been on static display at an airshow in the UK.

a planned debut to the public at what was to be the final RAF Waddington International Airshow just under two months later.

Thursday, 3rd July 2014 was the first of two arrival days for that year's show. The base, as it always did, opened up the grassed area at the north west of the runway as a Park and View enclosure for people to watch arrivals and display validations. To the delight of the assembled photographers and enthusiasts there on that day, 664 started up and taxied out past the enclosure, giving all those present a very good look at the RAF's latest acquisition. The aircraft then flew both days of the airshow weekend to participate in the traditional station flypast to open the show, along with Sentry and Sentinel aircraft. Around a week after the airshow, 51 Squadron deployed their new RC-135W on its first operational mission, flying out of Al Udeid Air Base in the UAE.

The second aircraft to arrive was ZZ665 during 2015. As Waddington was closed for runway works, 665 was delivered to RAF Mildenhall, a USAFE base in Suffolk where the 55th Wing has held an almost continuous detachment of its white top aircraft under the 95th Reconnaissance Squadron, on 4th September 2015. The final aircraft was ZZ666; the serials of all three aircraft were chosen to align with the original three Nimrod R.1 aircraft which were XW664-666. However, there were some discussions in the Air Force as to whether ZZ666 was the right number to use. With Nimrod R.1 XW666 having been lost in 1995 on a test flight out of RAF Kinloss it was thought the number may be a bad omen, but it stuck nonetheless. Interestingly, the aircraft that was to be ZZ666 was still operating out of RAF Mildenhall with the 351st Refuelling Squadron, part of the 100th Air Refuelling wing as a KC-135R during 2013 and the first half of 2014, departing for Greenville for conversion around the same time as ZZ664 deployed to Al Udeid.

For 51 Squadron's centenary in 2016, ZZ664 was painted in special markings to commemorate the milestone, with a large goose and Union Flag on the tail. The goose dates to when the squadron was flying the Avro Anson – Goose in Latin is Anser, so a little bit of wordplay was used here. As the squadron was also a heavy bomber unit at the time, a heavy wild fowl was deemed appropriate and the squadron crest was approved in 1937 by King George VI. The specially marked aircraft is seen here making an approach to Waddington on 4th September 2017 using the callsign 'Rooster 11'. As well as the Nimrods already mentioned earlier in the book, 51 Squadron have flown many types in their history, including the Handley Page Hastings, English Electric Canberra, de Havilland Comet, Sopwith Camel and Avro 504K.

ZZ666 seen on approach to Waddington on a bright summer's day, 28th August 2019.

Ascot 7224 RC-135W ZZ665 on final approach to Waddington's Runway 02 after a Black Sea mission on 24th May 2023.

ZZ665 'Shiner 50' on approach to Waddington being flown by crews from L3 Greenville, using the company callsign prefix of Shiner on 13th May 2021.

As previously mentioned, ZZ666 was the last of the three RC-135Ws to be delivered, again to RAF Mildenhall, on 6th June 2017. It was then flown onwards to Waddington the following day. This brought the total number of Rivet Joints in existence to 20 – Rivet Joints 1-17 belonging to the US Air Force and then 18, 19 and 20 – the three 51 Squadron examples.

The UK is the only country bar the United States that operates the Rivet Joint aircraft, which is down to the often-mentioned special relationship between the two countries. The fleet are all kitted out to an identical level now, enabling crews from both the US and UK to swap between aircraft should the need arise. It is also fairly common for RAF aircrew to join a USAF flight and vice versa should there be a gap in a mission crew.

A shock announcement came from the Pentagon on 8th January 2015 that no one could have ever imagined: the large USAF base at RAF Mildenhall, Suffolk, was to close by 2027. Mildenhall is a massive hub for aircraft staging out on deployment into Europe and further afield, as well as being home to the 351st Air Refuelling

Squadron KC-135Rs as part of the 100th Air Refuelling Wing and the 352nd Special Operations Wing, the latter flying CV-22 Osprey and MC-130J Commando II aircraft with the 7th and 67th Special Operations Squadrons, respectively. Mildenhall also had a permanent 'white top' unit – the 95th Reconnaissance Squadron with aircraft rotating from Offutt AFB on a regular basis since 1994.

The USAF has made no secret about its desire to centralise Rivet Joint ops in the UK and with Waddington ruled out due to its runway length, the US Defence Budget in 2017 had a sum of more than $30 million dollars set aside for the 95th to relocate to RAF Fairford, Gloucestershire. It was hoped that 51 Squadron would also move to Fairford, with centralised operations and ground support saving money for both the UK and US. Then summer 2020 arrived, Donald Trump was the President of the United States, the closure of Mildenhall was cancelled, and the sum of money set aside for the 95th to move was diverted elsewhere – most of it to aid the rebuild of Offutt Air Force Base, which was severely damaged by flood waters in March 2019.

Something that other aircraft stationed at Waddington hardly ever got was upgrades. Both the Sentinel and E-3D were retired early, partly due to being so out of date and behind similar NATO nations in spec. Thankfully this has not happened with the Rivet Joint fleet, with an almost constant upgrade programme that started with ZZ664. Little is known about the actual upgrades due to the secrecy that surrounds the aircraft and its systems, with an agreement with the US in place to keep the same level of classification around the Rivet Joint between the two countries. All that is known is that the aircraft have had a full glass cockpit upgrade.

RC-135s are often deployed where needed, with the most common being Souda Bay, Crete, to operate over the Middle East and in more recent years to monitor the situation in Russia and Ukraine. They regularly fly from their home base at RAF Waddington, meet up with a KC-135R tanker out of Mildenhall or Spangdahlem, and often the tanker uses the callsign 'Lager 51'. The missions with tanker support are normally around 10 hours, with sorties to the Baltic states or the Black Sea.

The Baltic missions could circle Kaliningrad, flying over the NATO nations of Estonia, Latvia, and Lithuania, or they could be flown up and down the eastern Polish border with the multitude of sensors gathering intelligence as they do. Shorter missions without tanker support can be around six hours and are generally to the Baltic states. The Black Sea missions became a lot more common since the Russian invasion of Ukraine and it is not unusual for the Russians to send up fighters to intercept the RC-135s just as RAF Quick Reaction Alert Typhoons intercept Russian bombers as they fly in the UK's areas of interest.

During a mission on 29th September 2022, RC-135W ZZ664, callsign 'Ascot 7215' was flying over the Black Sea in international air space on a typical intelligence gathering sortie. The Russians launched two Sukhoi Su-27 Flanker aircraft as they had often done – it was here that the similarities faded. The UK Defence Minister at the time, Ben Wallace, told MPs that an RAF RC-135 aircraft had been subject to a potentially dangerous incident, whereby one of the Su-27s had fired a missile in the vicinity of the 51 Squadron jet.

The UK were not treating the incident as a deliberate escalation by Russia, but took it as a reminder of the dangers faced daily by the UK military. Mr Wallace had spoken to his Russian counterpart and expressed his concerns and on 10th October, Russia advised that the outcome of the investigation concluded that there had

ZZ666, the last of the three to receive the glass cockpit upgrade, arrives back from the US on the morning of 11th September 2023 having originally departed Waddington on 9th June 2021. It was apparent on arrival that 666 had some additional upgrades alongside the glass cockpit with two additional sensors appearing on the upper rear fuselage. ZZ664 departed west on 14th September, presumably for the same upgrades using the same callsigns as 666 did — 'Same 01'. 'Same' being another L3 company callsign. The rotation system, with one aircraft always on long term maintenance or upgrade, was planned from the start. The RAF ordered three aircraft to have two available at any one time. There have been very few days when all three RJs have been on the ground at Waddington at once.

Ascot 7236 RC-135W ZZ665 returning to Waddington from an Intelligence gathering sortie on 26th February 2022, two days after Russia invaded Ukraine.

been a technical malfunction with the Flanker and its weapons. However, almost a year later, further information came to light that proved this was not the case.

On 14th September 2023, news agencies were reporting that the Russian pilot tried to shoot down the RC-135 after believing he had been given permission to fire. It appeared the Russians' rules of engagement and processes followed when asking for and receiving permission to fire are not in the same league as their NATO counterparts. The pilot fired the missile, but thankfully it did not lock on, and missed the aircraft.

A row then followed between the two Su-27 pilots with the second one saying he did not think permission to fire had been granted, despite this, the first pilot then released a second missile. It appeared that the weapon had malfunctioned or that the launch was aborted, as the missile simply fell off the wing hardpoint. Despite this incident, the RAF has continued to operate its RC-135s

A Ukrainian Air Force Su-27P in the flying display at the Royal International Air Tattoo 2019. The Su-27 is a formidable fighter aircraft, designed to compete with the McDonnell Douglas (now Boeing) F-15 Eagle and the now retired Grumman F-14 Tomcat. Russian Su-27s were seen at Waddington Airshow during the 90s, with the famous Russian Test Pilots putting on an awesome display in their specially painted red, white and blue aircraft.

over the Black Sea, albeit with armed Typhoon aircraft out of RAF Akrotiri, Cyprus, acting as fighter escort.

When operations allow, 51 Squadron partake in exercises both here in the UK, such as Cobra Warrior and Storm Warrior, as well as exercises throughout the world. In 2023, ZZ665 was away in the US for two months during the autumn, participating in Exercise Resolute Hunter 24-1 — an annual exercise focusing on C2 (Command and Control), ISR (Intelligence, Surveillance and Reconnaissance) and ELINT (Electronic Intelligence). Resolute Hunter is conducted from Naval Air Station Fallon, Nevada and brings together many assets and many different air arms which in the past has included the US Navy and their E-6B Mercury aircraft and the Royal Australian Air Force with their AP-3C Orion.

With the RAF working so closely with the USAF and their 55th Wing even before the RAF's Rivet Joint programme came in to existence, both sides have gained extensive knowledge and skills

ZZ664 cleaning up the gear after a touch and go at Waddington.

from working together on missions and during training. To save stress on the main mission fleet of their RC-135s, the United States also operates two TC-135W aircraft. These aircraft have the same elongated nose and cheeks on the front of the fuselage, but lack the multitude of systems of the mission aircraft, allowing pilots to train without having the need to take a front-line aircraft away from operational duties.

Due to the shape of the aircraft, it is not as aerodynamic as a 'normal' 135 series aircraft and has tighter restrictions on cross-wind and tailwind limits. The TC-135s have visited a handful of times since the RC-135s arrived at Waddington and as they are a rather rare aircraft, they always attract spotters. One such visit

Ascot 7227 RC-135W ZZ664 departing Waddington's runway 02 early on the morning of 27th March 2023, bound for eastern Europe — its light colours a stark contrast to the dark sky behind. MATT HALLAM

ZZ664 is seen here on a late summer's evening on taxi at RAF Waddington to participate in a Storm Warrior 23 sortie, using the 51 Squadron callsign 'DRAGNET 70'. The late summer's daylight catching on the multitude of aerials and antennae on both the top and bottom of the aircraft and giving it a golden glow.

TC-135W 62-4127 on taxi to depart RAF Waddington on 5th June, 2019 using the callsign 'Same 50'.

was between 3rd-5th June 2019 and with the TC fleet being some of the first to receive the glass cockpit upgrade, it was believed the aircraft was over to allow 51 Squadron personnel to learn more about the upgraded flight deck.

Another visit occurred in December 2022 using 'ANGUS 02' as the callsign, departing on the morning of the 9th. The next visit was a much longer and more active one and came during September 2023. The aircraft first arrived on Thursday 21st September but was only on the ground for around 25 minutes, picking up a handful of passengers and carrying out an engine running crew change before heading off to Malta Airshow for the weekend. The plan was for the aircraft to return to Waddington the following week and conduct some local training sorties after the conclusion of the airshow. It returned on the Monday after the show as expected, using the callsign 'Maverick 55'.

The TC then flew three local sorties that week, changing its

TC-135W 62-4127 'Maverick 55' on approach to Waddington on 21st September 2023 to pick up passengers for the Malta Airshow. The lack of mission antennae, sensors and aerials is very apparent here.

callsign to 'Maverick 51', the first of which was on the 27th when it went up to RAF Lossiemouth, Scotland, for a combination of radar and visual circuits. On the 28th the jet flew twice, firstly up to Lossiemouth for a few more circuits, then returned to carry out a hot pit refuel, something the RAF had not yet done with its RC-135s. A hot pit refuel is where an aircraft is refuelled with at least one engine running, sometimes involving a crew swap as well. It was on the ground for around 30 minutes, then took off for around 90 minutes of pattern work at Waddington, presumably with multiple pilots on, allowing them all to have an approach or two to the runway.

51 Squadron celebrated 10 years of operational service for the RC-135W on 23rd May 2024. A special flypast was arranged for Thursday 30th May, to coincide with a graduation parade at Royal Air Force College Cranwell. RC-135W ZZ666 callsign 'Goose 51' was joined by the Red Arrows, approaching Cranwell from the south. The large formation then carried on to Waddington, then over Lincoln Cathedral, followed by two more passes of Waddington before the formation split and landed. A great feat of airmanship with the two drastically different types flying close together for a prolonged period – a testament to the skill of both the RC-135W crew and the Reds' pilots.

Flight Lieutenant Dan Wilkes, 51 Squadron RC-135W pilot and organiser of this flypast provides a small insight into the preparation, planning and execution of the flypast: "Getting a larger aircraft at low level with fast jets off the wing takes a lot of planning and preparation; some five months of planning. Operationally, at height, formation flying with combat aircraft and various tankers is something we do often. But there is something different about flying 1,000ft off the deck with nine jets in a mixed formation off the wing. Leading the Reds is both an honour and a great responsibility.

"Through the crew of four flight deck members, we plan and fly the jet within the bounds of our strict rule set to ensure maximum safety and reward from the flight. The mixed formation flypasts that we have undertaken with the Reds form part of a larger celebration of 51 Squadron operating UK RJ for the last 10 years. It's both a privilege and a pleasure to be a part of the Royal Air Force's premier SIGINT squadron."

AIRFRAME HISTORY

ZZ664 – Delivered to the USAF on 6th November 1964, original number 64-14833 as a KC-135A.

Upgraded to KC-135R specification in 1991. Converted to RC-135W and delivered to Waddington 12th November 2013.

ZZ665 – Deliver to the USAF on 22nd December 1964, original number was 64-14838 as a KC-135A.

Upgraded to KC-135R specification in 1993. Converted to RC-135W and delivered to Mildenhall 4th September 2015.

ZZ666 – Delivered to the USAF on 20th October 1964, original number was 64-14830 as a KC-135A.

Upgraded to KC-135R specification in 1991. Converted to RC-135W and delivered to Mildenhall 6th June 2017, arrived Waddington 7th June 2017.

Close-up of the nose of ZZ664 showing the USAF serial of 64-14833, along with the City of Lincoln crest and the squadron commander's name at the time.

CHAPTER 9

ISTAR in a New Era: RPAS

RPAS — Remotely Piloted Air Systems — more commonly known as drones have been around for many years, but the RPAS in use in modern times are worlds away from the target drones such as the GAF Jindivik that served for around 50 years. The RAF's first RPAS was the General Atomics MQ-9A Reaper, which was developed from the Predator RPAS, which first flew in 1994. It takes three aircrew to operate a Reaper — the pilot, sensor operator and mission intelligence co-ordinator, who are often thousands of miles away from where the aircraft is, controlling it and its sensors via satellite communication from a ground communication system (GCS). The sensors include infra-red, daylight TV and targeting cameras, plus the forward-facing cameras used by the pilot for landing and taking off. The Reaper drone can carry weapons, with its usual weapons fit being two 500lb laser-guided bombs and four AGM-114 Hellfire missiles.

Reaper first flew in 2001 and entered RAF service in 2007, with the re-formed 39 Squadron operating out of Creech Air Force Base, Nevada with six aircraft. Operations began quickly with Reaper operating over Afghanistan. Sortie times could exceed 20 hours, allowing the RPAS to loiter over the battlefield providing live pictures thousands of miles away and able to engage the enemy should the need arise, which it often did.

A further five Reapers were ordered, which raised the need for an additional squadron to operate them — that squadron being 13 (XIII) Squadron, who had previously flown Tornado GR.4s. 13 Squadron re-formed at RAF Waddington in 2012 and continues to be the sole RAF squadron operating Reapers after 39 Squadron stood down in 2022. The Reaper drone is not certified to fly in all types of airspace, so has never flown in the UK and almost certainly never will — so photos of it outside of official military images are sparse. 13 Squadron operates Reaper daily, with several GCS units in the 13 Squadron hangar on base.

The next upgraded RPAS out of the General Atomics factory is the MQ-9B SkyGuardian, which in RAF service is known as the Protector RG.1. Protector is a larger, more complex drone able to carry a larger array of weapons and fly for over 30 hours at a time. Another major difference is that the aircraft is certified to fly in all types of airspace, meaning it can operate in UK and European skies. As such, the first SkyGuardian aircraft was able to be flown from Grand Forks, North Dakota in the United States, to RAF Fairford, UK to be on static display at the 2018 Royal International Air Tattoo. The RPAS flew over 3,700 nautical miles in just over 24 hours. It was wearing a civil registration, N190TC, but during the show the RAF announced that 31 Squadron were to reform to operate the Protector and as such the civil registration was removed and the aircraft had 31 Squadron fighter bars placed on the side of the fuselage and the squadron's gold star on the tail, as seen below.

The RAF announced that Protector was to be based at Waddington and a total of 16 were ordered, with an option for ten more. In order to test the airfield and its surrounding airspace N190TC returned to the UK and was operated from RAF Waddington and RAF Lossiemouth in August and September 2021. RAF Waddington and Lossiemouth were placed into a Temporary Danger Area (TDA), which could be activated when needed to allow the RPAS to operate safely.

The aircraft was painted in an overall white scheme, with its registration in large grey letters on the side of the fuselage. There was also an addition to the aircraft from the last time it was in the UK – it had a maritime radar under the rear fuselage as the manufacturer wanted to demonstrate the SkyGuardian's close relative – SeaGuardian. N190TC was first seen out in the open on the morning of 25th August 2021 conducting taxy trials all around Waddington, ensuring that both air traffic and the aircraft's operators were comfortable in their new surroundings.

SkyGuardian then conducted its first flight from Waddington later that same day and it continued to fly most days until departing to Lossiemouth. When flying, the RPAS would take off and circle over Waddington to gain the height required and once high enough it would be handed off to Swanwick Military Air Traffic Control to conduct its mission. The reverse happens on return to Waddington, arriving at height and descending inside the TDA to enable the aircraft to land.

The RAF received their first Protector – PR005 – in October 2022 some two years after the type's first flight. The aircraft has remained in the United States since delivery to assist in the initial training programme. The first Protector to arrive in the UK was PR009, which arrived just after sunrise on 30th September 2023 aboard an Antonov An-124 Transport aircraft. PR009 arrived in multiple pieces and would be assembled by personnel at Waddington in the coming weeks.

SkyGuardian taxying around RAF Waddington on the morning of 25th August 2021.

SNOOPERBASE WADDINGTON

On 2nd September, there was also a TV helicopter filming the RPAS, following it around the circuit and conducting flypasts along Waddington's runway as well as hovering just off final approach to catch the aircraft landing. DAVID MOORE

N190TC is seen here airborne from Waddington in September 2021. ANDY SHELTON

News of the An-124's arrival had spread, with General Atomics posting on 'X' (formerly Twitter) that the aircraft had departed the previous day from North Dakota. The assembled spotters all got their spot on the fence line to the south of the runway, but I decided to go for elsewhere, electing to go just north of the runway and use the wide-angle lens for something different.

Due to the sheer size and weight of the Antonov, it was not permitted to leave the runway surface, so after landing on runway 20, it turned around 180 degrees and parked on the 02 threshold to be unloaded. This took longer than expected due to an issue with the Antonov's nose, meaning everything had to be moved out of the back of the aircraft. There were no identifiable parts; pretty much everything was in boxes and once it was clear that these had to come out the rear of the aircraft, the assembled lorries all parked in a 'U' shape to further prevent people from seeing much.

On departure, the crew asked if they were able to depart off runway 02, so that they did not have to backtrack and turn around at the far end — despite a tailwind, which of course caused no issues for the rugged Ukrainian aircraft.

Protector RG.1 PR009 coded 'SM' was unveiled to the public on 23rd October 2023, in full 31 Squadron markings, 31 having reactivated at Waddington a few weeks earlier, although it would be 56 Squadron — the ISTAR test and evaluation squadron — that would be conducting the initial trials. It was first spotted outside a few days later, with ground runs being carried out in early November. PR009 is seen here towards the back to its hangar on 2nd November 2023.

The new danger area that covered both the Red Arrows and Protector went live on 1st November 2023, so it was hoped a first flight would follow soon after, but post rebuild issues delayed it for a few weeks. 16th November was the first time the Protector had moved under its own power; it was parked on the waterfront between hangars 4 and 5 and requested engine start using the 56 Squadron callsign 'Firebird 08'. It taxied down Alpha taxiway to the 02 end of the runway and then air traffic control fired a series of flares to check the Protector pilot could see them. Red and green flares are used for aircraft with radio failure to give them permission to land/take off or to abort, so it was vital that these were used. In the afternoon the same operation happened but this time for the 20 end of the runway.

17th November 2023, the Protector is parked out once again. It calls for engine start, this time using the callsign 'Firebird 09'. It was unclear whether this would be another taxy trial or a flight, but the aircraft entered the runway and was cleared for take-off, lifting off from the runway at 11:13 local. VIV PORTEOUS

XX295 lands after a training sortie on 8th March 2023, trailing smoke. Smoke is often trailed by the lead aircraft in the formation when they are about to touch down. This gives the following pilots an idea of how the wind might be affected by the aircraft in front; any of the pilots can request smoke to be trailed from the aircraft in front should the need arise.

favoured and that lasted until 31st March 2023. After that, the RAF lost control of the airfield and the airspace above it, so they could no longer use the location for display practices. It was hoped that an aerospace industry company would take on the base, allowing the team to continue using the sky over Scampton to practice, but that fell through when the Government announced it was to house asylum seekers.

There were several proposed airspace changes over Waddington with local general aviation pilots and stakeholders consulted. The result is a combined danger area known as EGR324 Alpha and Bravo over RAF Waddington for both the Red Arrows and the Protector drone operations. Up until this point the Red Arrows used to use a Restricted Airspace (Temporary) NOTAM – Notice to Airmen, whenever they needed to conduct a display practice

The Red Arrows arrive at Waddington as a six-ship after a short display over Scampton on the afternoon of 27th September 2022. The team now call the ex-Sentry dispersal, hangar and squadron buildings their home.

Each of the proposed new locations had its pros and cons and the local community hoped that the team would stay in bomber county. The RAF announced in May 2020 that the team were to move to RAF Waddington, but would still practice over RAF Scampton in the restricted airspace for as long as possible, they also started to practice over the east coast bombing range at RAF Donna Nook and the Central Gliding School at RAF Syerston, just south of Newark.

The latter two locations had their own issues, mainly down to safety with having no based fire cover, so Scampton was still

CHAPTER 10

Royal Air Force Aerobatic Team: The Red Arrows

Although a common sight in the skies over Lincolnshire for many years, the Red Arrows are one of the newer residents of RAF Waddington, having moved there in September 2022 from their previous base of RAF Scampton just north of Lincoln. They have been Lincolnshire residents since 1983, when they moved from RAF Kemble (now Cotswold Airport) and spent most of their time at Scampton, bar a period at the end of the 90s at RAF Cranwell to the south of Waddington.

The Cranwell move was down to defence cuts, with Scampton operations scaled down as part of the cuts. However, the Reds still practiced in the restricted airspace over Scampton, known as (Echo Golf) Romeo 313. Upon completion of each practice, a single Hawk would land at Scampton to collect the video recording of the sortie. The ground crew would store it in a small compartment in the nose and the aircraft would then depart back to Cranwell to join in the debrief with the rest of the team. With the Red Arrows needing sterile airspace, including when they took off and landed, it became apparent that they needed their own operating base again due to conflicts with the Cranwell-based training aircraft and so in 2000 they moved back to Scampton, where they remained for another 22 years.

With the announcement in 2018 that RAF Scampton was too close, three new potential new homes for the Reds were proposed – RAF Leeming, RAF Wittering and RAF Waddington.

ISTAR IN A NEW ERA: RPAS

Firebird 09 did one circuit and low approach over Waddington and landed at 11:28 after what was presumably a successful first flight. At the time of writing, Protector hasn't been airborne again but has been seen parked out multiple times. VIV PORTEOUS

RAF Waddington is going to be an international training hub for Protector, so we may end up seeing other countries' aircraft operating here at some point. They are a long way from the fast jets aviation enthusiasts have loved for a long time and a lot of people have said they won't even bother photographing them, but this is the way things are going and it's the next big development in the varied history of RAF Waddington and ISTAR. Let us see what the future holds!

The 2023 team conducting an ISP over Waddington on 11th August. The team were up to an eight ship in 2023 after the issues of the previous year that forced them to be a seven-ship and then a six-ship towards the end of the 2022 display season.

Winter training provides some different opportunities for photography, often with different display lines as the weather dictates, plus the low-lying sun can give some crisp winter light on the jets. Reds 1-7 plus Red 9 are seen here on the first eight-ship sortie of 2024 on 19th January.

A second shot from the first eight-ship of the 2024 season with the team in Big Vixen formation. Another bonus to the winter training is the chance to use the moon as part of the subject. Photographers can spend days trying to find the right spot for this; occasionally you get lucky and everything falls into place.

over Waddington. This was something they did quite a few times during the summer months to keep current, performing several In Season Practices (ISP) over the base.

The new 324 danger area became live on 1st December 2023 and is split into two – Alpha is a circle with a radius of five nautical miles centred on Waddington, up to 10,500ft. Bravo is a much larger rectangular block between 10,500ft and 19,500ft, which meets the controlled airspace just to the north of Waddington, with its most southerly point just above RAF Barkston Heath.

Winter training for the 2024 season started at the end of October 2023, with three new pilots and a new team leader – Squadron Leader Jon Bond. Jon brings a lot of experience to the team, having flown in the Red 2, Red 7, Red 6, and Red 4 positions as well as being the Tucano display pilot in 2012. 2024 marked the Red Arrows' 60th display season and the team will return to a nine-ship and be able to fly their signature formation – the Diamond Nine. When not flying as a nine ship, the team is split into two formations, Enid and Hanna. Enid are the front five, named after Enid Byton's Famous Five and the back four are now known as Hanna, after the late Ray Hanna who was a founding member of the team in the 1960s.

Hanna perform the more dynamic manoeuvres and are the more experienced pilots in the team and also includes the Synchro pair – Reds 6 and 7. During winter training, the formations are slowly built up, normally starting with Reds 1 to 3, with Reds 4 and 5 following once the initial three are comfortable. The Synchro pair work up starts with just one aircraft, with the new Red 7 in the front and the new Red 6 (last year's Red 7) in the back. Having watched a lot of the practices over the winter, it is great to see some old manoeuvres returning to the display, one of which hasn't been flown in over 30 years.

On the inside of the Vixen Bend.

It's 7th February 2024. Rumours are rife that today will see the team fly as a nine-ship. Nine Hawks are out on the line. Hanna calls up for the first sortie of the day, with Reds 6-9 flying their dynamic opposition crosses and dynamic breaks. The lunchtime slot approaches, Red 1 calls Waddington ground: "Good morning, Red Arrows requesting start for nine aircraft." The distinct sound of the Rolls-Royce Turbomeca Adour engines can be heard as they spool up and then the jets start their taxy.

Once all jets are in line, Red 9 calls over the team's frequency: "For the first time since October 2021 we are nine aircraft and we are all aboard." This signifies to the formation lead that all aircraft have left their parking spots and are on taxy together. For the first nine-ship of the season, the team practiced the first few large formation moves such as Vixen and Spitfire, with the third slot of the day also a nine-ship, the team practiced some of the more dynamic moves such as Tornado.

As mentioned previously, 2024 is the team's Diamond display season and a media day to officially launch the season and the special markings applied to the jets was planned for early March time. To aid in the season launch, a pair of Gnat aircraft arrived from North Weald on the afternoon of Thursday 7th March. The Gnat was the team's first mount from 1964-1979 before they moved on to the Hawk T.1. On the morning of 8th March, one of the Gnats flew with five of the present day Red Arrows around the local area, with Red 10 acting as photo chase. The formation flew over Lincoln Cathedral a few times then circled round for a flypast over RAF Waddington. The Gnat, Red 1 and Red 10 then split off for some one-on-one photos.

Tornado, a crowd pleaser since its arrival, returned in 2024, but from the left of the crowd rather than the right as it was in 2023. This move symbolises the start of the dynamic second half of the show.

One of the older moves returning for 2024 is the 5/4 split and cross, last flown in 1993. The team flew the move as a nine-ship three times during the final practice on 7th February. The team arrive from crowd right in Diamond Nine after completing Tornado, then split into Enid and Hanna while flying away from the crowd at 90 degrees (known as the B axis) before returning and crossing in front of the crowd.

A close-up image of the markings on the side of the fuselage and half way up the tail fin, a simple yet effective '60' with a pair of Hawks on either side, with their 'smoke' passing through the middle. Some of the public like the markings, others think they are a little on the small side — but they are as big as the current RAF legislation would allow!

The Gnat that flew with the Reds was in the scheme of a Yellow Jacks aircraft. The Yellow Jacks were one of the many predecessors to the Red Arrows, but were only around for one year after it was decided the RAF should only have one full time display team rather than the multiple trainer and front line squadrons that had popped up in the previous years, such as the Black Arrows and Blue Diamonds flying Hawker Hunters, the Red Pelicans flying Jet Provosts, and The Firebirds flying the massive English Electric Lightning.

A couple of close-up photos of the Gnat, showing Red 4 in the back seat to act as a liaison between the Gnat pilot and the Reds. Red 4, Flt Lt Ollie Suckling, also flies classic jets including the Gnat, which explains why he was chosen for the support role for the sortie. The Gnat also had some of the team's 60th Anniversary markings applied to the tail prior to the flight, but they did not last long, with only a small amount left at the top of the tail!

The Gnats taxy to depart late in the day on 8th March. ANDY SHELTON

Some more shots from winter training...

With the traditional Diamond Nine formation returning, it's only right that the team arrive from crowd rear in a full diamond.

The diamond appeaed twice in the 2024 display, the second time being at the start of the second half when the team split into Enid and Hanna.

Red 7 Flt Lt Tom Hansford pulls up to start a barrel roll in the Synchro Opposition Barrel Roll cross.

Red 9 Flt Lt Patrick Kershaw exits crowd right after the Hanna pass.

Pheonix. One of the larger formations for the first half of the display.

Tango. Reds 1-5 in line abreast. Even numbers are on the leader's right, odd are on the left.

The famous Concorde is once again replicated by the team for the 2024 season.

The Red Arrows ground crew are known as the Blues and once PDA (Public Display Authority) has been approved they all wear bright blue overalls where the pilots are allowed to wear their traditional red flying suits. There is also a select team of Blues who are lucky enough to fly with the pilots; these are known as 'the Circus'. The Circus travel in the back seats of Reds 1-10 (and 11 when needed) when the team are forward deploying to another operating base or display location. Circus 1 is normally the JEngo (Junior Engineering Officer), Circus 10 is normally a photographer who captures the amazing air-to-air shots and Circus 11 is normally the SEngo (Senior Engineering Officer).

The remaining members of Circus are all from various other trades, such as avionics technicians, mechanical technicians and armourers. Opposite are some images of the Blues working on the jets and assisting with the start up and taxy out process. Each jet has one dedicated ground crew member who assists the pilot with strapping in; they then marshal the jets out of the parking locations and on their way. There are also two additional groundcrew, who walk around all jets and then walk to the end of the flight line and do one final check for any leaks or open panels as the jets taxy past.

Red 3 Flt Lt Dustin Wales performs a walkaround of his jet and checks the ailerons.

Red 8 Flt Lt Richard Walker climbs aboard his jet.

Red 9 Flt Lt Patrick Kershaw sits in his jet, engines running, with the Blues member on the right assisting with the flight control systems functionality checks.

Reds 7, 8 and 9 taxy out of the parking bays with a final thumbs up from the ground crew — all good and ready to fly!

Shots from the flightline...

CHAPTER 11

Exercises and Deployments

Over the past 40 years, Waddington has had several exercises with aircraft from across the globe descending on the Lincolnshire base, mainly due to its proximity to the east coast air combat ranges and vast parking spots left over from the V-bomber days. As this book is all about the base since the arrival of ISTAR with the Nimrod AEW, we look at the variety of aircraft, of which the majority are fast jets, that have operated from Waddington during exercises since the mid-1980s.

During the Cold War years, mass deployments to Europe by United States based aircraft were common. One such deployment to Waddington was by the United States Air National Guard with aircraft from three separate A-7 Corsair units — the 175th Tactical Fighter Squadron from South Dakota, the 174th Tactical Fighter Squadron from Sioux Falls, and the 124th Tactical Fighter Squadron who were from Des Moines. A total of 36 aircraft arrived at Waddington on 11th May 1985, with 12 jets from each squadron — 11 single seat A-7Ds and one twin seater A-7K from each.

During the exercise, the squadrons conducted simulated attacks into mainland Europe, which included low level flying and occasionally a visit to an air combat range on the continent. They would also land at USAFE bases in Germany or Holland to refuel before returning to Waddington, bases visited included Spangdahlem, Hahn and Soesterberg. The jets also spent some time at other NATO bases in Europe to practice air combat manoeuvres with different types and air forces, such as the Dutch F-16s out of Leeuwarden and the West German F-4F Phantoms out of Wittmundhafen.

During their time here, the Corsairs also visited some of the east coast bombing ranges, including RAF Wainfleet and RAF Holbeach, known as 'the Wash' ranges, with Wainfleet on the north side and Holbeach on the south. Holbeach is still active today, but Wainfleet

A-7K 81-0077 'IA' from the 124th Tactical Fighter Squadron of the Iowa Air National Guard. KEN WITHERS

is now a unique holiday destination — its tower a six-bedroomed accommodation with 360-degree views from the visual control room, which is now the living room. Also available to sleep in are a retired RAF Jetstream T.1 aircraft and Army Lynx AH.7 helicopter — I doubt the A-7 pilots could have ever imagined that! The A-7 could carry 500lb and 1,000lb 'dumb' bombs, the AGM-65 Maverick air-to-ground missile and the AIM-9 Sidewinder for self-defence, but during range visits practice bombs were used.

These were much smaller and only had a small amount of explosives in, to create a small flash for the range operators to identify

An A-7D from the 124th TFS in a wraparound camo paint best suited to the colours of Europe. KEN WITHERS

how close to the target the weapon landed. Another benefit to the pilots was the weather variations in Europe, allowing them to fly and train in much lower visibility and temperatures than they would get back home. The exercise ended on 8th June 1985, with the aircraft departing to their home bases after a successful few weeks in the UK and Europe.

Quite possibly the largest gathering at Waddington was for an exercise that occurred in the summer of 1986. Tactical Fighter Meet attracted a whopping 70 aircraft from the RAF, USAF and NATO countries. The list of participants is very impressive and the variety

Fighters as far as the eye can see. A Tactical Fighter Meet package on Delta taxiway, led by F-16s from Nellis AFB. Also visible are RAF Tornados, more F-16s and then bringing up the rear four RAF Jaguars. On the flight line to the left are another Tornado and some USAF F-15C Eagles. GRAHAM ROBSON

An FB-111A of the 509th Bomb Wing at Pease Air Force Base, New Hampshire. The FB-111s differed from the F-111s based at RAF Lakenheath and RAF Upper Heyford, so were a welcome sight for the UK enthusiast. The FB-111 had strengthened undercarriage, a wider wingspan and was around 2ft longer than a standard F-111. GRAHAM ROBSON

of types is something we can only dream of today.

The RAF provided Phantom FGR.2s from 19(F) and 56(F) squadrons, Tornado F.2s from 229 OCU (Operational Conversion Unit), Tornado GR.1s from 17 and 31 Squadrons and Jaguar GR.1/T.2s from 54 Squadron. The USAF provided F-15C Eagles, F-16C Fighting Falcons and FB-111As. From European nations there were Mirages, F-16s, NF-5s, Tornados, Drakens, F-4 Phantoms, European-based CF-188s and a sole NATO E-3A.

The aircraft operated in mass missions, that were specifically briefed with the departures conducted in waves. Prior to the exercise, the participants had pre-exercise training, where they were paired up – air combat aircraft with 'mud movers': low level strike aircraft. The pre-exercise training was done in late July, with the aircraft heading to Waddington upon completion. There was also a photo call on the Saturday before the event kicked off, with the public invited on to Waddington to view the multitude of aircraft, with at least one from each unit being on static display on the western side of the airfield.

The main exercise only lasted a week and started on 4th August, emphasising the importance of the pre-exercise training between various types so they could benefit from day one. Some pilots and crews flew as many as five missions that week, the air defence/fighter aircraft flew various profiles such as a Combat Air Patrol (CAP) – flying round in a pre-set area waiting for targets, supported by tankers and airborne early warning aircraft, with missions lasting up to three hours. Other missions would be fighter escort, providing support to the attack aircraft, flying ahead of them to clear the way and spot the enemy sooner, keeping them busy to allow the attack aircraft to press on to the target.

The exercise came to a crescendo on the final day, with three 16-ship ground-attack waves escorted by multiple air defence fighters – the target was to be RAF Spadeadam, an electronic warfare range in the middle of Cumbria. All the aircraft departed from Waddington in a 50-minute window, a near constant stream of jets lining the taxi ways and powering away down the runway. On the Friday afternoon, the participating crews were treated to a short flying display from the Battle of Britain Memorial Flight and the Red Arrows from nearby RAF Coningsby and RAF Scampton, respectively.

ACMI – THE GLORY DAYS
British Aerospace (now BAE Systems) opened the Air Combat Manoeuvring Instrumentation facility – ACMI for short – in August 1990. This consisted of four units at RAF Lakenheath, Leewarden AB in Holland, RAF Coningsby and RAF Waddington, with the latter being the main hub for the operation and an air combat range out over the North Sea, about 80 miles off the east coast.

The range consisted of six towers anchored to the sea bed known as Tracking Instrumentation Sub-System towers, with five of them arranged in a circle 30 nautical miles in diameter, with the sixth in the centre. These towers tracked the aircraft during their missions, with each participating aircraft carrying an ACMI pod under its wing. The pod looked like an AIM-9 Sidewinder missile without the fins and a spike-like pitot tube on the nose. Each of the four ground units had the ability to play back the missions to allow the pilots to debrief and learn from each one. The screens at the main control centre at Waddington could display 36 aircraft at any given time, but BAe were able to load as many aircraft as they had pods available – with aircraft quantities in the low forties not unheard of.

The first deployment to Waddington from an overseas nation to use the range was the Belgian Air Component, who arrived in October 1990 with three F-16s. The Belgians were one of the most common visitors to Waddington to use the range, deploying multiple times a year throughout its existence, often alongside Armee de l'Air (French Air Force) Mirage 2000s. As well as the regular ACMI visits, larger exercises started to occur, such as the Swiss Air Force Nordseekampagne (Norka) exercise which ran for several years in the 1990s and the large Exercise Nomad, which ran from the late 90s until the range closed.

The range's demise started around three years prior to closure, with USAFE pulling out of the agreement and the debriefing facility at Lakenheath pulled down. Advances in technology also played their part in the closure – new GPS-based ACMI systems allowed the aircraft to operate 'range-less' without the need for fixed towers to record aircraft positions. Mobile phones were also gaining popularity and would interfere with the range's signals; both factors made the range less popular and with fewer customers year on year, BAe could no longer operate it and make a profit. The range closed in 2005 and there was a final Nomad planned for April 2005. It was hoped this would be the range's swansong, but it was sadly cancelled.

The sign outside the ACMI complex at RAF Waddington. GRAHAM ROBSON

Exercise Nomad was the largest of the ACMI exercises and usually occurred during July, with the first edition being held in 1996 and the final in 2004. The ACMI regulars of Belgian F-16s and French Mirage 2000s were joined by various other nations such as the Spanish Air Force with their EF-18 Hornets and RAF participation came from Tornado F.3s. The Swiss also joined in 2000, deciding to merge their Norka exercise with the much larger Nomad to allow them to gain more experience. The Norka and Nomad that followed allowed the Swiss to stretch their legs, so to speak, with their home airspace heavily restricted, with no flying below 1,000ft and supersonic flying banned.

USAF F-15C Eagle 80-0035/IS from the 57th Fighter Interceptor Squadron, based at Keflavik, Iceland taxies down Delta to depart on an ACMI mission in October 1991. Note the conformal fuel tanks fitted on the side of the engine intakes to extend the range of the jets — a must for the squadron's remote operating location. GRAHAM ROBSON

Spanish Armada AV-8A Matador VA.1-5 01-806 from 008 Escuadrilla taxies back to parking after an ACMI sortie in June 1992. This aircraft was part of a batch sold to the Thai Navy in 1996 and is now on display at the Royal Thai Navy Museum, Samut Prakan, Thailand. KEN WITHERS

June 1993 saw another edition of the Swiss Air Force's NORKA exercise, with Mirage IIISs and F-5E Tigers operating from Waddington. Seen here is Mirage IIIS J-2038 with his wingman taxying behind. KEN WITHERS

EXERCISES AND DEPLOYMENTS

Austria visited with their Saab 35OE Drakens to use the range a few times, their first visit being in 1995 with jets from 2 Staffel and the final in 2002, when they brought nine aircraft with the aim to fly six per day against Harrier GR.7s from RAF Cottesmore. Draken 22 is seen here on taxy for the active runway in May 1995. The small tailwheels at the back of the aircraft can be seen here, to prevent the long tail of the aircraft scraping on the ground during landing and take-off rolls. KEN WITHERS

Luftwaffe MiG-29 Fulcrum 29+12 from JG73 'Steinhoff' taxies back after an ACMI sortie after streaming its braking parachute in July 1997. The MiGs were originally part of the East German Air Force that were inherited by the Luftwaffe after the reunification of Germany. The MiGs were made as NATO compatible as possible and provided an excellent adversary aircraft for NATO members to practice air combat with. GRAHAM ROBSON

A typical Nomad view from 2004. Four Armee de l'Air Mirage 2000s taxy down Delta, with 94/12-YI from EC1/12 'Cambresis' leading the way. The French always parked at the end of the taxiway at a 45° angle which gave a good side on view of the lead aircraft with a head on of the rest as they approached the photographers at the fence line. On departure, the French conducted a Hi-Lo take-off when the weather allowed — the first and third jets doing a very steep climb out with the second and fourth keeping it low.

A lineup of six Swiss Air Force F-18 Hornets parked on Delta dispersal for the final Nomad in 2004. The final occurrence was smaller than the previous ones, with sadly no RAF or Belgian aircraft operating from Waddington. The final ACMI deployment came towards the end of 2004 with the usual suspects of French Mirage 2000s and Belgian F-16s, bringing an end to one of the most fondly remembered periods of Waddington's history.

EXERCISES AND DEPLOYMENTS

Su-30MKI SB 042 taxis to out for a morning sortie on 4th July 2007, visable in the background are two of that year's airshow participants — Hunter T.7 G-VETA and Canberra B.2 G-BVWC (ex-WK163 — a record breaker).

After the closure of the ACMI range in 2004 and with it the demise of Exercise Nomad, Waddington became a little quiet, bar a small detachment of Marine Nationale (French Navy) Super Etendards in 2005 that was heavily curtailed by weather, but that all changed for a few weeks in 2007. Exercise Indradhanush – meaning Rainbow in Hindi, has now been held five times – the first of which was in India in 2006. Waddington was the first UK base to host the exercise in 2007, with Coningsby hosting it the second time in 2015.

In 2007, the Indian Air Force sent six Sukhoi Su-30MKI Flankers to Waddington. This was the first time the IAF had deployed the

25 Squadron Tornado F.3 ZE728 'FZ' and Su-30MKI SB 044 lined up on the runway at Waddington ready to depart on 4th July 2007.

type overseas and the first time they'd deployed to the UK. The jets were from 30 Squadron of the IAF, known as the Rhinos, and they were supported by IL-76MD transport aircraft from 44 Squadron and IL-78 MKI Tankers from 78 Squadron. The RAF participation was in the form of Tornado F.3s from 25 Squadron, based at RAF Leeming as well as the still relatively new Typhoons of 3 Squadron, which operated from their home base of RAF Coningsby.

The exercise was all about air combat and air defence, with the sorties being flown out over the North Sea. These started off as 1v1 sorties with a Tornado flying against a Flanker and slowly building it up to a mass formation mission with a mix of Su-30s, Typhoons and Tornados. The Flankers with their canard wings and vectored thrust made it very hard work for the much older F.3s – the Tornados flying in a 'clean' fit (no wing tanks) to aid their manoeuvrability in the air. Russia was not happy that the Flankers were participating in a UK-based exercise and as such the Su-30 pilots were prevented from using their highly classified N011M Bars radar system during the exercise for fear of anything being picked up by the USAF Rivet Joint aircraft from RAF Mildenhall.

If the pattern continues, the Indians could be back in the UK for the sixth instalment by 2025

ENTER THE SNAKE
After the excitement of Indradhanush in 2007, it once again became a little quieter at Waddington. The RAF had been holding its Combined Qualified Weapons Instructor course, which was the largest exercise the RAF ran, with aircraft generally operating out of RAF Kinloss near Inverness, for several years. The course was one of the final exams for personnel to gain the Qualified Weapons Instructor status.

Although being held at Kinloss, the exercise was run by the Air Warfare Centre which is based at Waddington. This exercise become what we now know as Cobra Warrior, which is run by Waddington-based 92 Squadron, the RAF tactics and training squadron. The name for the exercise is derived from the Cobra that is at the centre of 92 Squadron's crest. Usual participants from the UK would be RAF Typhoons and F-35Bs supported by Voyager tankers. The USAF would send F-15s and F-35s supported by KC-135 tankers as well as the occasional transport aircraft from Brize Norton. Helicopters are often included in the exercise too, although they mainly operate from further north, such as RAF Leeming.

The first Cobra Warrior was in March 2016, with foreign participation being provided by the German Luftwaffe, with EF-2000s from TLG71 and TLG73 being based at nearby RAF Coningsby. The second instalment in September 2017 saw things start to get a little busier, with the EF-2000s from TLG73 returning plus some Tornado ECRs from TLG51, again operating out of RAF Coningsby.

September 2018 saw the first Cobra Warrior participants at Waddington, with four Italian Air Force F-2000A Typhoons from 4 Stormo being based here, parking at the very southern end of the base. Of course, the Germans participated again, this time with 16 aircraft – eight EF-2000s and eight Tornados at RAF Coningsby and the USAF sent F-16s from Aviano Air Force Base in Italy to RAF Lakenheath.

Cobra Warrior 2019 again occurred in September and it went mad. Parking restrictions appeared around Waddington, laybys were closed, security was heightened as the base geared up for the expected hundreds if not thousands of spotters all wanting to get a glimpse of this year's participants, all of which were based at Waddington.

The Italians were there with their F-2000s, with four aircraft from a mix of three Wings, ten Luftwaffe EF-2000s from TLG73 and TLG74 and for the first time the Israelis were participating. They brought over a mix of F-15C and F-15D 'Baz' aircraft from 106 Squadron, supported by KC-707 Re'em tankers from 120 Squadron.

The Israelis brought a new level to Cobra Warrior and the fence line at Waddington was full every single day with enthusiasts making long journeys to come and see these rare aircraft. The Israelis also brought their own Mossad security, with additional measures put outside the fence where the F-15s were parked at the southern end. Mossad were also spotted on the first morning's sortie as the jets taxi'd past the photographers on Delta – that was the only time they came that route, on all subsequent missions the Israelis taxied down Alpha taxy way on the other side of the runway.

Although this is a shot from Coningsby, it's included as it helps tell the Cobra Warrior story. This is TLG51 Tornado ECR 46+57 on taxy for runway 07 at Coningsby on 27th September 2017, preparing to depart for home upon completion of Cobra Warrior 2017. The aircraft is carrying two practice AGM-88 HARM (Highspeed Anti-Radiation) missiles under its fuselage.

EXERCISES AND DEPLOYMENTS

36 Stormo (Wing) F-2000A MM7353 comes in to land after a sortie on 9th September 2019.

Taktisches Luftwaffengeschwader (Tactical Air Force Wing) 73 'Steinhoff' EF-2000 30+50 enters the runway on 11th September 2019.

F-15C 'Baz' 818 from 106 Squadron parked on the southern bays by Crash Gate 5 on the evening of 31st August 2019. Note the one and half Syrian Air Force kill markings on the nose.

The Israeli contingent all departed a few days prior to the exercise ending. The first to set off was F-15D 'Baz' 715 performing a rather nice climbing banking turn over the A15 to meet up with the tanker that had already departed.

During 2020 and 2021, the Covid years, the exercise did run, but on a much smaller scale with all aircraft operating from their home bases. Hopes were up as the RAF announced Cobra Warrior would be run twice a year, in March and September and with a new type to the UK expected for Cobra Warrior 22-1 excitement was building. The Indian Air Force were to attend with their HAL Tejas, sending five aircraft. Also planned to attend were Belgian F-16s and Saudi Typhoons, with the first aircraft expected to arrive by the end of February 2022.

Due to Russia's invasion in Ukraine, the exercise was cancelled just over a week before on 25th February, which although disappointing to lots was completely understandable. Cobra Warrior

Watching the KC-707 Re'em tankers depart was rather spectacular — they were using Runway 02, so taking off towards the A15 and the WAVE. They both used near enough the whole length of the runway and lifted off with less than 1,000ft of the runway to go, obviously heavily laden with fuel to get the F-15s home.

22-2 did go ahead as planned in September. The Luftwaffe returned with Tornado ECRs from TLG51 and the Italians returned with F-2000 Typhoons from 4, 36, 37 and 51 Stormos.

The Italians also sent a new type to Waddington for Cobra Warrior 22-2, a E-550 CAEW (Conformal Airborne Early Warning) aircraft from 14 Stormo. The jet was still needed on operations by the Italian Air Force and as the exercise only ran every other day, the jet flew home, conducted its operational sortie, then returned to Waddington in time for the next exercise. It did this throughout the whole two-week period. The USAF also participated with some

Rage 11 flight roll to the end of runway 02 after a CW 22-2 sortie on 15th September 2022.

non-UK based aircraft, sending F-16s from Aviano again.

Cobra Warrior 23-1 was set to be a big one. Rumours we rife as to what would attend, with confirmation only really gained when a copy of the exercise patch appeared. The patch depicted the Belgian F-16, Finnish F-18, Turkish F-16 and KC-135 Tanker, Saudi Typhoon and Indian Air Force Mirage 2000. Sadly, due to an earthquake in Turkey in early February, they elected to withdraw from the exercise to allow their military to provide support where it was needed. A rumour also appeared online that the Indians had cancelled, thankfully this was a misunderstanding, with someone reading the press release from 12 months earlier when India had confirmed that CW22-1 was indeed cancelled.

The Saudis were the first to arrive, with an A330 MRTT (Multi Role Tanker Transport) arriving at Waddington on 26th February – a feat in itself – thanks to the hard work of the base operations team as this was the first time an A330 aircraft had landed at Waddington. The aircraft was full of personnel who were transported by road to nearby RAF Coningsby where the Typhoons were

E-550 CAEW MM62303 14-12, callsign 'Perseo 71' exits the runway on 15th September 2019.

A pair of Mirage 2000TIs taxy out for a sortie on 23rd March 2023, with the full contingent of Finnish F-18 Hornets parked on the lazy in the background.

to operate from, with the Typhoons only coming to Waddington once for the press day. The Belgians and Finns tended to fly twice a day, with the Indians flying once most days. The snow in the second week brought an unexpected element to the exercise and the Finns felt quite at home.

Cobra Warrior 23-1 saw the first time in many years that Waddington hosted a photo call event. It was the 'Cobra Warrior enthusiasts event', with only 50 free tickets available, but with a suggested donation of £50. The tickets had all been snapped up within 30 minutes of being advertised. Sadly, the first attempt was cancelled due to the snow, the base deeming it unsafe for people to travel long distances to see the aircraft.

This Indian Air Force Mirage 2000TI KT211 was on show at the photo event. The crews were very friendly and happy to answer any questions about their aircraft.

Finnish Hornet HN-422 at the photo event on 21st March.

Thankfully, the base rescheduled and the event went ahead as planned on the final week of the exercise, just on a smaller scale as the Belgians had returned home at the end of the second week. Fifty enthusiasts boarded a coach in the officers' mess car park and were driven to bays 1-9, which is where the Red Arrows are now based, just in time to see the synchro pair taxy back to parking.

They then spent the next hour photographing a Finnish F-18 Hornet and an Indian Mirage 2000TI, both nations providing aircrew for people to speak to and the Indians brought a lot of memorabilia, with some people spending more than £150 on badges, caps, prints and T-shirts.

A PARTICIPANT'S THOUGHTS

In February/March 2023, we embarked on an exhilarating journey across the North Sea to participate in Cobra Warrior 23, a monumental exercise held in the United Kingdom. We were based at RAF Waddington.

Cobra Warrior itself was a dynamic challenge, pushing our F-16s to new heights alongside international partners. One of the highlights of our visit was the opportunity to explore the iconic Mach Loop in Wales. The breathtaking scenery and challenging low-level flying routes left an indelible impression on our pilots. The Mach Loop not only showcased the natural beauty of Wales but also served as a true test of our skills and precision.

We were particularly delighted by the enthusiastic presence of numerous spotters, who gathered to capture the essence of our F-16s in action. Their passion for aviation photography added an extra layer of excitement to the entire experience and we are grateful for the stunning images they captured, immortalizing our participation in Cobra Warrior.

Our time at RAF Waddington and participation in Cobra Warrior was an unforgettable chapter in the history of the 349 Fighter Squadron. The lessons learned, the friendships forged and the memories created during this exercise will undoubtedly resonate within our squadron for years to come.

We extend our gratitude to the Royal Air Force and all involved for a successful and enjoyable exercise.

Sgt Bram Thieren, 349 Squadron, Belgian Air Force, 10th Tactical Wing, Kleine-Brogel Air Base

During 2023, word came about that the second running of Cobra Warrior that year would be a little more exclusive. As such, Cobra Warrior 23-2 was a smaller affair with Waddington hosting seven CF-188 Hornets from the Royal Canadian Air Force's 433 Squadron, based at Bagotville and a pair of NATO E-3A AWACS aircraft.

The Italian E-550 also returned, but was only around for the first week due to operational needs. Norway provided some of its F-35As, which operated from RAF Lakenheath, and the week prior to the exercise a Norwegian F-35A was part of a three-ship from Lakenheath that conducted a practice diversion to Waddington. In addition to the UK based American participants, the US also sent some F-16s from Spangdahlem Air Base in Germany and they operated out of RAF Mildenhall — their two-seater F-16DJ was often seen hurtling through the Lake District at low level.

This exercise was a little more intense, with exercise sorties flown every day, rather than every second day as in previous years and there was also a night flying element on the third week which added something new for the enthusiasts. You would be surprised how many people were on the side of the A15 on a dark cold September night trying to photograph Hornets!

The 349 Squadron patch created for the exercise. BRAM THIEREN

F-16AM 'The Duke' of 349 Squadron taxies in after arriving at RAF Waddington on 1st March 2023. BRAM THIEREN

EXERCISES AND DEPLOYMENTS

Three Belgian Air Component F-16AMs taxy out using the callsign 'Mace flight', with FA-77 'Mace 12' in the lead on 16th March.

Finnish Air Force PC-12 PI-05 on taxy to depart at the end of the exercise.

Cobra Warrior 23-1 attracted a lot of support flights — 47 in total, with 21 of those just from the Saudis. The Finns had a couple of private charter flights, plus a Learjet and two PC-12s, the Indians had C-17s and an IL-78MKI tanker, the Belgians had multiple A400 visits and the Saudis had C-130 and A330 movements to Waddington, Coningsby, and East Midlands airport.

Saudi A330 MRTT 2403 on approach to runway 02 to collect the Saudi personnel on 26th March.

Indian Air Force IL-78MKI KJ-3453 is prepped for departure on 2nd March.

Indian Air Force C-17 Globemaster CB-8004 arriving on the Saturday morning after the exercise finished.

CF-188 Hornet 188735 on approach to runway 20, the jet still retaining its scheme from when it was the 2021 Canadian Air Force solo display jet.

2023S BONUS

To the delight of the locals, Cobra Warrior was not the last of the fast jet deployments for 2023 — Waddington was due to host Exercise Atlantic Trident 23-1. Atlantic Trident is an exercise between the RAF, USAF, French Navy (Marine Nationale) and the French Air Force (Armee de l'Air). The base was expecting to host four Armee de L'Air Rafales as well as four Marine Nationale Rafale Ms, but the latter chose to operate from their home base. Waddington also hosted a French E-3F AWACS. It was bizarre to see an E-3 in the skies over Waddington the same month that the remaining four RAF E-3s were scrapped.

French E-3F 204/36-CD departing Waddington on 3rd November, bound for Brize Norton for a practice diversion which ended up being denied by Brize Air Traffic Control.

Part way through the exercise all the participants landed away at RAF Leeming, to give the air and ground crew the chance to operate from an unknown base with limited support. The four Air Force Rafales accompanied by four RAF 12 Squadron Typhoon FGR.4s, two RAF F-35Bs and four USAF 495th Fighter Squadron F-35As landed at Leeming after the sortie on 3rd November, the French Navy jets would land there on 6th November. All aircraft returned home on Tuesday 7th November, with the exercise finishing on 10th November.

This was the first time Rafales had visited RAF Waddington. As per the Mirage 2000s in the ACMI days, the jets parked at 45 degrees to the taxiway and departed in a Hi-Lo stream when the weather allowed. Boston 21 flight are seen here on taxy just before a very heavy shower on 8th November.

CHAPTER 12

Air Traffic Control Recollections

RAF Waddington ACMI Exercises — by retired Sergeant Max Shortley

Air Combat Manoeuvring Instrumentation or ACMI as it was known — four letters guaranteed to instil a healthy mix of nervous excitement and respect into the mind of a recently arrived air traffic controller at RAF Waddington.

I arrived at RAF Waddington in October 1998 following a lengthy five-year tour at RAF Newton, the RAF's last all-grass airfield for fixed wing powered flying where typical trade consisted of Chipmunks for Air Cadet AEF, Bulldogs for the East Midlands University Air Squadron and T-67 Firefly aircraft for the Joint Elementary Flying Training School (JEFTS). Their fixed undercarriage and very sedate speed around the circuit, did little to prepare me for the enhanced pace of air traffic control at a main operating base like Waddington, which regularly hosted multinational fast jet exercises making use of one of the few European ACMI ranges.

British Aerospace (BAe) had their facility on the eastern side of the airfield to host these exercises. The infrastructure used comprised a series of rigs out in the North Sea in an area designated as EG D316 and EG D317, meaning 'EG' for the UK and 'D' for Danger Area. Detachment fast jets would on arrival be fitted with telemetry pods for the missions; these then communicated with sensors on the rigs so that combat manoeuvres between opponents, where participating squadrons would take turns at playing red and blue forces, could be watched at the BAe facility in real time. Exercise sorties were also recorded for later playback and analysis as part of the mission debrief to the participants once they had landed.

Max in the Visual Control room at Waddington. Directly in front is a laminated map of the airfield and visual pattern, where controllers use small markers to identify where an aircraft is in the pattern. Upright directly behind that is the airfield traffic light system. To the right, the red crash phone and a CCTV screen looking on to Alpha dispersal, which was obscured from the VCR by the large Alpha hangar. All images in this chapter, unless otherwise noted: MAX SHORTLEY

Fortunately for me, I wasn't a newcomer when it came to controlling fast jets, having previously spent three years at RAF Marham in the early 90s with three squadrons of GR.1/GR.1A Tornado aircraft. For my swansong shortly before leaving, I was privileged to be entrusted with the role of RAF ATC Liaison Officer situated in the control tower at the Italian air base at Decimomannu in Sardinia for a 617 Squadron ACMI detachment making use of the Capo Frasca range. Our jets were joined by a detachment of RN Sea Harriers providing the Dissimilar Air Combat Training (DACT). A week later they were all joined by another detachment of Tornados from 14 Squadron based in Germany. So, I had a half decent idea as to what to expect at Waddington.

Arriving in October '98 I had only recently missed the last ACMI exercise of that year and as I had not seen a radar screen for over five years, I first had to attend a mandatory two-week terminal refresher course at RAF Shawbury to refresh on unused ATC skills. Back at Waddington and with the arrival briefs out of the way, I got stuck straight into on-the-job training.

As the Local Examining Officer (LEO) was a fellow controller who I had worked with at Marham, she rightly expected me to pick things up very quickly and get up to solo validation standard in double quick time. Endorsements in Talkdown and Zone which provided a Lower Airspace Radar Service (LARS) to civil and military aircraft transiting Waddington airspace followed before the festive break. These endorsements were quickly followed by Ground and Aerodrome Control (ADC) in January '99. It wasn't long before my numbers came up for the obligatory four-month detachment 8,000 miles away in the windswept South Atlantic!

My official detachment dates were February to May 2000 but once you're on the Detachment Warning Roll (DWR), if anyone drops out ahead of you, everyone shuffles forward. I was required to qualify as a Radar Director before proceeding to the Falklands—this control position had caused me a few sleepless nights during my initial controller training at Shawbury and then again during training at Marham. So I knew this was a validation I wasn't going to get either easily or quickly!

The year marched on and other controllers kept being put

AIR TRAFFIC CONTROL RECOLLECTIONS

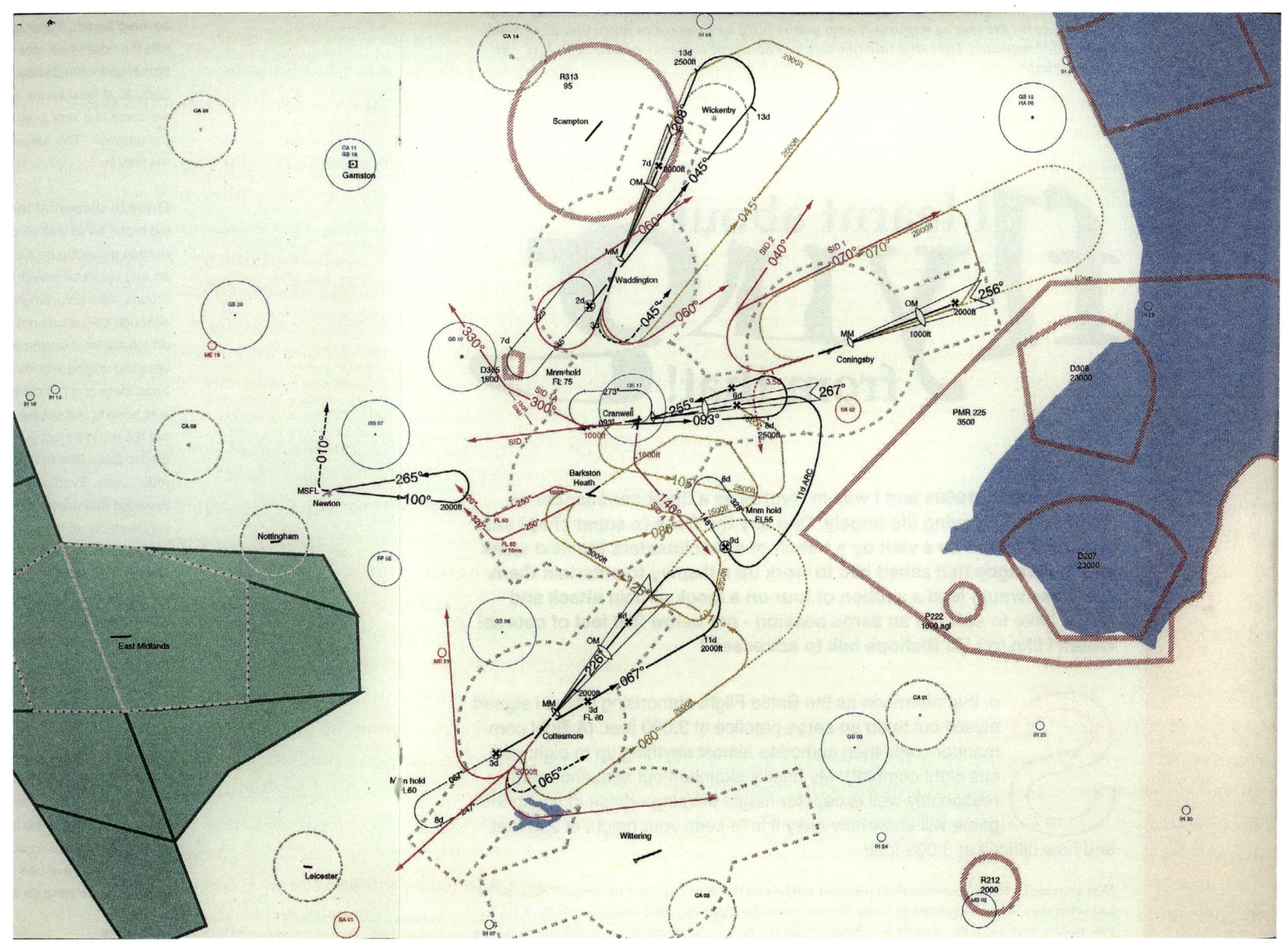

A map showing the various intertwined radar patterns for the multitude of bases in Lincolnshire.

ahead of me in the training queue. Eventually it was recognised that I needed some priority and Director training began in earnest. There had previously been a rule that only Approach-qualified controllers could radar direct for ACMI inbound traffic due to the complexity of dealing with mixed packages of multinational fast jets, with pilots speaking English in a multitude of dialects and some barely at all. Fortunately, this rule was rescinded and I trained over several weeks with whatever traffic was generated and this included ACMI traffic.

Departing ACMI traffic en route to the exercise area would almost exclusively depart on one of the published Standard Instrument Departures (SIDs). These would always be handled by a Radar Approach-qualified controller as Waddington ATC didn't have a separate 'Departures' console. The name of the game though was to get them established on the SID, with London Mil already prenoted of their departure and an SSR squawk pre-allocated. The Approach Controller would aim to have the aircraft identified immediately off the end of the runway and the SSR mode 'Charlie' verified and handed over to London Mil as soon as possible once established in the climb and clear of any conflictions, usually before passing 10,000ft.

An age-old adage for RAF controllers was always to get rid quickly and not hang onto traffic unnecessarily, this principle was always applied to ACMI traffic, as class G airspace in Lincolnshire was well known as 'bandit country' where conflicting traffic either had to be avoided by standard separation minima or co-ordinated, all of which could very easily tie an already busy controller in knots.

The direct to console London Military phoneline during an ACMI mission would be answered by the ATC Supervisor who would make the call on how each recovery wave would be controlled. They would also act as Allocator when the time came to take the handovers from London Mil. The reason being that the workload would be too great normally for a single radar control position to handle all the inbounds.

So the recovering fast jets would be split sequentially between the Approach Controller and the Radar Director. Factors such as runway selection and prevailing weather conditions determined whether this would be a Visual Flight Rules (VFR) or an Instrument Flight Rules (IFR) recovery. If the weather conditions were marginal, any early participants to head back to base, ahead of the main recovery wave, where the outcome of a successful or aborted VFR landing could set the precedent for the remainder, but often this was just pre-determined by the ATC Supervisor and IFR recoveries were the order of the day as this wasn't something to gamble with.

On the larger ACMI exercises, up to 20 fast jets would recover to the base in short order. The departure wave would have involved large 'packages' as they were referred to, and getting them all off to the exercise area was the relatively straightforward part. The recovery wave was determined largely by the order in which London Mil handed them over to us; aircraft could return as singletons, in pairs or threes etc. Individual aircraft endurance was a factor, so any declaring a fuel priority could be handled accordingly. Pilots that had been 'shot down' early on would usually be in the first elements to recover, so we knew their day hadn't gone particularly well!

Fast jet pilots with the mission/exercise complete will aim to get back on the ground as quickly as possible, the preferred option being a visual run in and break. If the weather conditions were suitable, they would be vectored by the radar controller to the Initials Point (IP) for the runway in use, these being 4nm out

The old radar approach room at RAF Waddington, showing the various positions a controller could work. The radar controllers for Waddington are now based at RAF Coningsby, working in a new Air Traffic Hub known as the Lincolnshire Terminal Air Traffic Control Centre, which opened in 2022.

in the approach and displaced by 1nm on the dead side. Radar would descend them to 1,500ft on the airfield QFE and once they had called "visual" with the airfield they would chop across to the tower frequency to join. The ADC would clear them to join down to 1,000ft on the airfield QFE having called the position of any other traffic already in the visual circuit.

They would then run in on the dead side and break from one third to two thirds along the runway length onto the downwind leg on the live side. Excess speed would be rapidly bled off in the turn with the use of airbrake and then the gear selected down to increase drag. It was straightforward to get a lot of aircraft on the ground in the shortest possible time when VFR recoveries were possible. Sadly all too often the weather had other ideas and the cloudbase and visibility would determine that recovering aircraft had to make IFR approaches, either PAR or ILS for runway 21.

However, the fun would really begin when runway 03 was in use, with a talkdown as the only precision IFR option as there wasn't and still isn't an ILS installed on Runway 02 as it is now. Radar controllers needed to really up their game to maintain an orderly and expeditious flow of recoveries. It became essential to have both talkdown positions running simultaneously with ideal track separation between the individual aircraft or pairs on recovery kept to 4-5 miles apart. Aircraft in emergency or having declared low on fuel would be afforded priority whenever possible, but often they were simply fed in as expeditiously as possible without throwing a major spanner in the carefully choreographed order of recovery!

Having spent three years controlling at Marham, I had already discovered that fast jets have a tendency to have more than their fair share of emergencies and the fact that ACMI aircraft were mostly from foreign air arms made no difference to that equation!

My favourite seat in the house was upstairs in the Visual Control Room (VCR) as the Aerodrome Controller (ADC). Here I was at my most comfortable and I revelled in wheeling and dealing, mixed packages of fast jets both departing and arriving. In the radar room, you were very much a member of a close-knit team of controllers sat either side of you and with the ATC Supervisor hovering over your shoulder. But upstairs in the VCR you were pretty much king of all you surveyed.

The ATC Supervisor couldn't be in two places at once, so unless an emergency was in progress, there was usually just me and the VCR Assistant and the liaison with the radar controllers was through the headset. Sometimes we would be joined by one of the

The visual control room during the 2010 RAF Waddington Airshow. The VCR was always much busier during the airshow days!

detachment pilots, who could provide advice to us if we needed it, but also to speak on the radio to their own aircrews if they had a problem and needed advice. They had their own desk with comms at the back of the VCR and some good, friendly banter was often the order of the day!

I recall one F-18 Hornet rolling down the runway and as it passed just abeam ATC in my one o'clock, I was drawn to what appeared to be a cloud of feathers engulfing the aircraft, my immediate thought was of a birdstrike. But the eyes can play tricks with you, as the aircraft had actually dropped its chaff dispenser on the runway and this had emptied its contents on impact, so what I was actually seeing was a cloud of small foil pieces billowing around. Chaff was a defensive measure designed to fool radar guided missiles fired at the aircraft. That took some clearing up as obviously we had to take Foreign Object Debris (FOD) very seriously!

The two control towers at Waddington where Max spent much of his time. On the left, the 1950s control tower and on the right the newer control tower which was opened in 1996. MIKE HILEY

Without doubt though, the most memorable incident occurred on 4th November 2002. I had a large 'package' of 11 French Mirage 2000s to depart to the exercise area. Runway 20 was in use and the aircraft were cleared to depart in pairs in a ten second stream. This would have been one of the larger formation departures during any ACMI exercise and I was giving due consideration to the responsibilities resting on the formation leader. In RAF terms, leading a formation is a qualification that is awarded based on merit and experience, the larger the formation, the more experienced and qualified the pilot has to be.

I cleared them all to take-off and watched as the first pair commenced their take-off roll, the pair slightly staggered using both sides of the 200ft wide runway. As they were abeam the tower about 2,500ft along the runway, I noted that the leader was no longer keeping pace with his wingman, and all too quickly the aircraft had come to an abrupt halt and was now blocking the eastern half of the runway a little over one third along the runway length. In 2002, RAF controllers were still permitted to issue a verbal "abort" instruction to pilots taking-off, so with the next pair already commencing their own take-off roll, I wasted no time getting on the radio to issue an abort instruction. My assistant and I were working together to coordinate the necessary alerting of the crash line crew as a high-speed abort was always taken very seriously.

One Mirage, the leader's wingman had safely got airborne and I quickly sent him to Approach Control for the SID departure. That was the only Mirage 2000 from that departure wave that succeeded in launching, the jet had to burn off fuel to get down to landing weight anyway before attempting a landing, so working with the Approach Controller was the best course of action. The nine remaining aircraft all taxied back to dispersal via the runway, in 'elephant walk' style. A team of aircraft engineers then joined the fire crew on the runway at the stricken Mirage to assess and make good the retrieval of the aircraft. ATC had by now dispatched the Deputy SATCO to establish what had happened.

Visually from the VCR using binoculars it appeared that the aircraft had embedded itself into the runway surface, which definitely had me scratching my head and I envisaged some fairly major runway damage! It transpired, however, that the pilot, who happened to be the French squadron boss, had forgotten to take

All images on this page: ROGER STEELE

the parking brake off prior to commencing his take-off roll, the sheer power of the engine in reheat had made the aircraft slide along the runway with the wheels not rotating.

The tyres duly burst and the alloy wheels then started to grind down on the rough friction surface of the runway. So, by the time the aircraft had come to a halt, the main wheels were now almost ground down to the hubs, a very embarrassing 'game over' for the pilot!

During my Radar Director training I recall one session where we were on runway 03 with IFR recoveries in force. We had a large mixed package to recover from the ACMI range. As usual the elements were handed over from London Mil to myself and our Approach Controller in turn. Very careful airspace management for sequencing was required on these occasions. The first to recover from the north east would be pointed at the airfield and once clear of the Belmont TV mast and closer to the airfield, descended to 2,000ft through the overhead for a teardrop pattern onto the extended runway centreline for a PAR approach.

Subsequent elements would be routed increasingly westerly before turning onto a south westerly heading. I don't recall if we had both talkdown positions available, possibly not, so separating the formations into singletons and pairs and sequencing them ready for talkdown was going to be a challenge in order to keep the spacing between elements as tight as possible. Various unknown radar returns had to be avoided and other known traffic coordinated with adjacent units. Consequently, I found the airspace required to keep these elements a safe distance apart was getting wider and wider and I was now rapidly running out of airspace in which to vector my singletons and pairs.

I was instructed to pick up the landline from East Midlands Radar; the curt response from the other end simply said, "Much as it's nice to see your fast jets, can you get them (expletive!) out of my radar pattern..." Click. The line went dead! It really wasn't a great training session, but I certainly learnt about sequencing, airspace management and careful vectoring from that comment!

I recall during another ACMI exercise involving RN Sea Harriers, the detachment commander had repeatedly expressed concern about the state of the surface of the parallel eastern taxiway (now 'D' taxiway), to the extent that he had the poor junior ratings out every morning before flying commenced, hand-sweeping large sections of taxiway to mitigate any FOD ingestion as, by all accounts, the Sea Harriers' engines had been rather energetically hoovering up any loose particles of grit and this was causing damage to their compressor blades.

Inevitably, with so many different nations participating in ACMI exercises, language barriers were bound to crop up from time to time. In ATC we often found that in formations of three or four fast jets, the leader would be fluent, but his wingmen might only have a very limited knowledge of English. A strict requirement for us in aerodrome control was to obtain a verbal acknowledgement from each pilot on visual final approach that the gear was down. A feature of the French Air Force fast jets was that they were fitted with a tone when they transmitted with the gear down, we knew it as the 'Roger Bleep.'

So, a typical situation in the visual circuit as each called final, we would say, "Confirm gear down?" and in response we would get "Bleep." After a couple of rounds of this, we would then say "I need you to say that your gear is down!" We had tried to get dispensation from our external Standards Evaluation Unit to be able to accept the 'Roger Bleep' but they wouldn't give it! Those old enough may

A pair of Royal Navy Sea Harrier FA.2s taxy down delta for an ACMI sortie in 2004. MIKE HILEY

recall that back in the day when the RAF trained their fast jet pilots on the Jet Provost, these too had once been fitted with the 'Roger Bleep' to mitigate against student pilots experiencing a high cockpit workload interfering with that essential safety element of not trying to land or roll with the gear up! But getting your 'ear in' on what some of the pilots transmitted was an acquired skill that quickly developed as each exercise progressed.

For several of the participating foreign pilots, the UK weather must have come as something of a shock as ACMI exercises weren't confined to just the warmer summer months! I recall one ACMI when we had Spanish F-18s detached to us, one was arriving solo. I believe it might have been an airframe swap for another unserviceable jet. We were on runway 03 and we were experiencing low cloud and poor visibility.

The pilot had elected to make a visual landing and I'd cleared him to land but couldn't see the aircraft due to the low cloud. The pilot couldn't see the runway either, despite the lights being on maximum brilliancy, so It looked like a missed approach situation where I would have to send the aircraft back out to radar for a PAR approach. However, at the last possible moment, the pilot had obviously been able to see the runway and by now the aircraft was just moving into view from out of the grey sky gloom over the eastern parallel taxiway abeam the PAR (Talkdown radar) at a height of no more than 200ft, in a right turn with a fair degree of angle of bank on. Half the available runway was now behind the pilot, but skilfully and akin to a carrier deck landing, the pilot firmly planted the jet down and quickly brought it under speed control. That was one of many ACMI moments where my hand had been reaching for the red crash phone!

French Air Traffic Control has been infamous for decades for the frequent strikes that seem to afflict the service. We were not immune to these problems, as our cousins across the Channel always seemed very quick to announce "Non" when we tried to obtain clearances for jets returning home after a detachment! At the end of one ACMI, we had a large package of French Mirages all backed up on the eastern taxiway for a runway 20 departure, but the all important 'French ATC Acceptance' wasn't forthcoming.

After some quite considerable delay and with most of the cockpit canopies now popped open to prevent the pilots overheating, the formation leader dug his mobile phone out and called someone back home, we have no idea who, or what strings he pulled, but basically he obtained the French ATC clearance himself! Bizarrely, I have even been aware of the same problem post airshow when we've had participation from the Patrouille de France, their own national team!

Waddington hosted ACMI exercises five or six times a year with the biggest and longest exercise being 'Nomad'. These lasted several weeks, so members of the ATC squadron were well versed in the 'work hard, play hard' mentality. There would usually be a good end of exercise party thrown by one of the squadrons and if the detachment had gone well and they had liked us, we usually got invited!

The relationship between aircrews and ATC was always convivial; at the very least we could expect to get a nice framed and signed squadron print to hang in the corridors. But I think that a Swiss F-18 squadron took the honours as the best party hosts during my time handling ACMI traffic. I cannot help wondering though if the fact that some of the female spotters that lined up at the hedge line on the A15 and had been lifting their tops up to give the pilots an eyeful had any bearing on them enjoying their time at RAF Waddington?

Sunrise over Waddington, taken from the balcony of the VCR, looking east. Visible are three of the retired Sentinel aircraft.

CHAPTER 13

Visitors and Diversions

Due to its location, and vast parking bays, Waddington has had its fair share of visitors over the years. Front line aircraft would often drop in for a fuel stop or just a practice approach after hitting any one of the east coast ranges. Training aircraft from Europe would conduct a navigation exercise to Waddington, stop for lunch and return home, then there were the RAF training aircraft. This of course does not occur as much these days due to fewer aircraft being around, but Waddington does still get a varied mix of visitors, with the past few years being bumper years thanks to the Station Ops team making Waddington more desirable and available. If we look at 2023 alone, I photographed 61 different types at Waddington. There will be a few extras that I did not catch, but that is still some going.

RAF training aircraft are the most frequent visitors to Waddington. It is a rare day when something from Cranwell or Wittering has not visited the circuit. Sadly, the types visiting are far less varied than they once were, Jetstreams and Dominies were replaced by King Airs and now the King Airs have been replaced by Phenoms.

The Boeing E-7A Wedgetail has visited Waddington three times up until now, with all visitors coming from the Australian Air Force. The first visit was in 2006, with A30-005 being on static display for the airshow, the second (pictured, page 204) was A30-006 which visited in 2017 with some rather nice nose artwork and the third being A30-001 which was there for a few days in 2018 after participating at RIAT at RAF Fairford for the RAF100 celebrations. The Australians have six of these aircraft, so Waddington has done very well for half the fleet to visit.

The E-7A has been purchased by the RAF to replace the E-3D and it was originally planned that the E-7s would operate from Waddington, with a new purpose-built facility on the eastern side

We will start off in 1990, with this EC-24 from the US Navy. The aircraft was first delivered to United Airlines in September 1966 as a DC-8-50 and ended its life with the US Navy, being transferred to them in 1987. The aircraft arrived at Waddington from Volkel Air Base, on 25th September 1990 as 'Nucar 03'. 163050 was withdrawn from use in 1998 and is now in storage at Davis-Monthan AFB. GRAHAM ROBSON

of the runway. All the plans were drawn up and things were looking positive, but it was announced in December 2020 that the small fleet would instead be based at RAF Lossiemouth, which is also home to the RAF's P-8 Poseidon fleet – a very similar aircraft, with both platforms being based on the Boeing 737 airliner.

14th December 2022 saw a mass diversion of aircraft from the 48th Fighter Wing at RAF Lakenheath, with four F-35As and seven F-15Es diverting in due to fog at Lakenheath and their primary diversion base. Most of the jets managed to depart around 4pm the same day, with pretty much all of them performing a quick climb departure – keeping it low along the runway and pulling up sharply with the afterburner glowing, which certainly made for some impressive photos in the crisp winter twilight.

14th September 1993 and a Polish Tupolev Tu-154 was at Waddington, to transport the exhumated body of a Polish Home Army soldier killed in Gibraltar in 1943 during the Second World War. The soldier in question was General Wladyslaw Sikorski, who was buried in Newark Cemetery. A large memorial still stands in the cemetery where the General's body was for 50 years. GRAHAM ROBSON

SNOOPERBASE WADDINGTON

Graham Robson recalls a divert of U-2R 'Clip 37' on 10th April 1994: "The U-2 was a weather diversion on a Sunday afternoon; whenever at home I always used to listen to London Mil on my scanner and on this day, I heard 'Clip 37' asking for the weather at his home base of RAF Alconbury, but the crosswind was out of limits. London Mil advised him Waddington was the only suitable diversion base. When the U-2 changed to Waddington approach, I was convinced there was a chase plane flying with it as another callsign was giving constant info updates to the U-2, in addition to what Waddington were saying. It was vectoring to runway 03 as it was back then, and I was at the Harmston end before it landed. As I was waiting, I heard a fast car approaching and eventually it came into view — it was a USAF chase-car roaring up the A607. This was who had been talking to the pilot, the car constantly updating him on his position as the car was needed to be on the runway when it landed as a guide and then to attach the pogo wheels. I photographed it land, then raced up to the A15 and luckily it was still on the runway when I got there." GRAHAM ROBSON

In the summer of 2004, for a short period of time, Waddington hosted a USAF E-3 Sentry from Tinker Air Force Base, Oklahoma, along with a KC-135 Tanker from the 157th Air Refuelling Wing. The aircraft are seen here on taxy on 4th July — the E-3 coming from Alpha dispersal with the RAF E-3Ds in the background and the KC-135 coming down Delta taxiway, with a thumbs up from the pilot.

Clockwise from the top left — 207 Squadron Tucano T.1 ZF139, 45 Squadron Beechcraft B200GT Super King Air ZK460, 55 Squadron Dominie T.1 at the 2006 Waddington Airshow and finally a 45 Squadron Phenom T.1.

NATO E-3 Sentries were once a very common sight both on the ground and in the circuit at Waddington; now they are a lot less frequent, although we did have three examples supporting both Cobra Warrior exercises in 2023. E-3A LX N-90446 is seen here on climb-out after one of many approaches on 27th March 2023 as 'NATO 40' — a common E-3 training callsign.

Another common NATO type from the 90s and early 2000s was the Boeing 707TCA — Trainer Cargo Aircraft, known as a CT-49A. The last of these aircraft were retired in 2011. LX N-20000 is seen here on approach to Waddington to participate in the 2007 Waddington Airshow

VISITORS AND DIVERSIONS

Three images from 2008, the first from 17th June, parked at the southern end of the airfield was US Navy P-3C Orion 161005 from the USN Test and Evaluation Squadron — VX-1, based at Naval Air Station Patuxent River, Maryland. The aircraft served for another six years after this Waddington visit, before heading to the 309th Aerospace Maintenance and Regeneration Group (AMARG) — also known as 'The Boneyard'.

French Air Force E-3F 201 taxies down Delta on 17th September, with the aircraft sporting 70th Anniversary markings.

Another Airborne Early Warning aircraft that occasionally visited Waddington is the French Navy (Marine Nationale) E-2C Hawkeye, the small fleet of three aircraft is operated by Flottille 4F, with this example seen on 9th October. The French have ordered three new E-2D models to replace their E-2C Hawkeye 2000s with deliveries expected to be completed by 2028. All images: KEN WITHERS

A civilian-operated IL-76 from Silway Air Cargo, based in Azerbijan arriving on 25th February 2010. KEN WITHERS

USAF Air Mobility Command C-5M Galaxy 14th April 2011. KEN WITHERS

A nice little Italian Air Force P.180 Avanti, taken on 2nd October 2013. The Avanti is an interesting little aircraft — it has pusher propellers mounted on the large rear wings and smaller canards at the front, it almost looks like it is flying backwards when one passes over you. KEN WITHERS

What could have been... One of the visits of an Australian E-7 in 2018. KEN WITHERS

On 5th February 2019, four 41 Squadron Typhoon FGR.4 aircraft from nearby RAF Coningsby diverted in due to poor weather shortly before 11am. They all departed just before 3pm. Here they are on taxy down Delta as a four-ship using the callsign 'Apollo 11' flight.

The rarest visitor in recent times was this Royal Saudi Air Force RE-3A Tactical Airborne Surveillance System (TASS) which arrived on 30th September 2021. Aircraft 1901 is the sole example of this aircraft type and it is seen here finally departing late on 1st October after multiple issues with its flight plan. It was en route to the United States for rework and upgrades by L3 in Greenville.

RAF 'heavies' do occasionally visit RAF Waddington to conduct practice approaches and circuits, the most common being the P-8 Poseidons of 120, 201 and now 42 Squadron at RAF Lossiemouth. ZP801 'Guernsey 03' is seen on approach to runway 20 on 27th September 2022.

October 2022 saw the national display team of Switzerland — the Patrouille Suisse, operating out of Waddington for their display at Duxford's autumn airshow. The seven F-5E Tiger II aircraft arrived on 6th October and had hoped to conduct a practice over Duxford on the 7th, but the weather was atrocious. They ended up practicing in the morning, then returning in the afternoon to conduct their display. They are seen here on taxy for their display on the afternoon of 8th December. DANIEL KENNEDY

One of the diverted jets was the 48th Fighter Wing heritage marked F-15E. It is seen here rolling to the end of Runway 02. ANDY SHELTON

494th Fighter Squadron F-15E 91-0309 lights the afterburners of its Pratt & Whitney engines and roars down the runway. MAX SHORTLEY

Homesick Angel. An F-35A climbs away in the dusk sky. GAVIN TURNER

Each year, the flying assets of the UK military perform a flypast for the monarch's birthday or other major anniversaries, such as the 100th Anniversary of the Royal Air Force in 2018 or the King's Coronation in 2023. There is always a practice and for a number of years the practice has been conducted using College Hall Officers Mess, at nearby RAFC Cranwell, just to the south. 2023 was no different in that respect, however, the rotary element operated from RAF Waddington for the practice on 25th April. Helicopters from the RAF, Royal Navy and Army Air Corps descended on Waddington. We had four Juno HT.1, two Apache AH-64Es, three Wildcats, one Merlin, one Chinook and one Puma. All the aircraft lined up on runway 02 and departed in formation as 'Spectre Combine' at 13:58 and returned 35 minutes later after a successful practice. 'Lifter formation' are seen here arriving on the morning of the flypast – Chinook HC.6 ZH897 leads Puma HC.2 XW235/Q.

In late spring 2023, rumours started to circulate that the Royal Saudi Air Force Saudi Hawks Display team were coming to Waddington, prior to them participating in the Royal International Air Tattoo at RAF Fairford. The team arrived in glorious sunshine on Tuesday 4th July, using the callsign 'Baz 01' along with a support C-130H Hercules. The team flew on Friday 7th July with the Red Arrows. The Reds displayed first, then the Saudis departed and met up with them out over the Lincolnshire Wolds. The 16 aircraft formation conducted a flypast over the Humber Bridge, then Lincoln and the International Bomber Command Centre. The Reds then landed to allow the Saudis to put on their show. The Saudis performed for a second time the following day, during RAF Waddington's families day.

In August 2023 some of Brize Norton's 'Heavies' were operating from Waddington for a few weeks while Brize underwent runway repairs. This saw a number of A400 and C-17 flights in and out of Waddington, although not as many as originally planned due to a couple of unserviceability issues with the C-17 fleet. 99 Squadron C-17 ZZ175 'Ascot 6649' is seen on final to runway 02 on the evening of 17th August having flown from Bahrain.

During the first week of Cobra Warrior 23-2, Waddington hosted a small number of Hawk T.2 aircraft, from 4 Squadron who are based at RAF Valley, with the aircraft conducting night flying. 'Reptile 31' flight are seen here on taxy in the last light of the day on 4th September 2023.

VISITORS AND DIVERSIONS

French E-121 Xingus are the most common visitor to Waddington these days, some weeks they visit as many as four times, bringing French E-3F crews to use the E-3 Simulator at RAF Waddington. Xingu 078/YE is seen here on 2nd October 2023 sporting a very recently applied scheme to celebrate the 40th anniversary of the Xingu in French Air Force Service.

CHAPTER 14

Visiting ISTAR Assets

For a period, RAF Waddington hosted some Britten-Norman Islander and Defender aircraft of the RAF and the Army Air Corps. The Islanders and Defenders had been around since the mid-1960s and had many ISTAR and covert roles, their low speed and excellent endurance making them an ideal platform for loitering for long periods of time. They tended not to hang around, often landing long (further down the runway) and departing from the midpoint, avoiding the spotters on the fence line.

The type was retired from military service in 2021, but some of them do still operate in civilian hands, still conducting a similar role. One was conducting some sort of surveillance operations near Waddington in February 2024; the aircraft landed at the base for a fuel stop using the callsign 'Bandit 08'. The Islander was G-BVSK, operated by GAMA Aviation – the crew were not keen on being photographed and didn't hang about – expediting their taxy to the runway.

RAF Islander CC.2 on the runway at Waddington, 29th March 2017. Hard to see from this image but the rear of the aircraft's cabin was blanked out.

An unknown Defender on final to Waddington, 11th October 2017, the rear of its cabin blanked out. Also viewable are multiple aerials along the fuselage and wing and a large camera on the nose, most likely a WESCAM MX-15.

Army Air Corps Defender DL.2 ZG996 comes in to land on 24th February 2011 in a different sensor fit to the previous two aircraft.

This unusual aircraft was at Waddington for a few days in 2019. At a first glance it just appears to be a civil Piper PA-31 Navajo but with the addition of a camera under the nose. The aircraft is now owned by 2Excel Aviation LTD and operated on to HM Coastguard.

CHAPTER 15

Waddington International Airshow

There is already a very concise book on the Waddington International Airshow by Gary Parsons, so I won't go into great detail here, but it feels right to include it as it was a big part of the base for almost 20 years. Waddington had held airshows and photo calls before, but 1995 saw the first incarnation of what became one of the highlights on the airshow calendar right from the off. Up until 1994, RAF Finningley held an airshow on the traditional Battle of Britain weekend in September as so many RAF bases had done, but with the base winding down and eventually closing in 1996 a new location was needed and Waddington was chosen.

The first airshow was held on 1st-2nd July 1995, with the star of the show being the USAF F-117A stealth fighter. There was an example in the static display and for the first time in the UK, an example in the flying display. The inaugural show also had the Russian Test Pilots in the Su-27 Flankers performing an awesome pairs display. Other highlights included the Israeli F-15i Ra'am display in 2001, multiple visits by the B-1B Lancer and the return to the airshow circuit of Avro Vulcan XH558 in 2008. Waddington did well with display teams too, including the Frecce Tricolori from Italy, Sarang — the Indian Air Force helicopter display team and the USAF aerial demonstration, the Thunderbirds, who visited with their immaculate F-16s in 2000, 2011 and 2014.

The 2014 show looked to be a good one and it was. The Red Arrows were celebrating their 50th display season, the powerful 'Solo Turk' F-16 display was over, the Swedish Air Force historic flight had sent their Draken and Viggen solo displays and the USAF Thunderbirds were back for a third and final time. It was known there was to be no show in 2015 and possibly 2016 due to the planned works to extend the runway and remove the 'hump'

F-117A 82-0805 lands at Waddington on one of the days prior to the show in June 1995 — the airshow team certainly got off to a good start securing these as participants. GRAHAM ROBSON

The show in 2000 managed to attract a pair of Republic of Singapore Air Force Skyhawks. A single seat A-4SU and this twin seat TA-4SU were on static display for the weekend. The aircraft were on long-term deployment to Cazaux Air Base, France as part of an advanced jet training detachment.

In 2001 the Israeli Air Force sent three of their F-15i Ra'am aircraft to Waddington, supported by two KC-707 tankers and a C-130 Hercules, with the Hercules remaining on static display for the airshow. The F-15is were from 69 Squadron — 'The Hammers' and as the Israeli military is very sensitive about displaying its aircraft, the fact that they were participating was kept very low key, with information provided on a need-to-know basis only. As such there was no public announcement prior to the show that they were participating. The F-15i is very similar to the US F-15E Strike Eagle and jet number '269' flew a couple of times on both days of the show, flying more of a combat demonstration than a display.

in the middle of the runway, mainly for RC-135 operations, but the airshow was cancelled in 2015.

The show was then cancelled permanently over increased security concerns — a decision probably influenced by the USAF, with one of their most secretive aircraft now being based at Waddington. Another probable factor was drones, with the Reaper MQ-9A being controlled from Waddington since 2013. The blow was softened with the possibility of an airshow at nearby RAF Scampton, but that would not happen until 2017. Scampton did indeed have a show in September 2017 that was run by the same people who do the Royal International Air Tattoo, but it wasn't on the same level as either RIAT or Waddington's show that it was intended to replace.

Both the flying display and static display were a little weak, meaning not many people attended. There was also very limited parking on base, with most having to park at the nearby Lincolnshire Showground and be bussed in.

2004 saw the first public display of the Eurofighter Typhoon, flown by RAF pilots. A pair of examples from 17 Squadron, who were still based at BAE Warton at the time, were present, with one static and one flying example. Both jets are seen here on taxy to depart on the Monday after the show. They both departed in style it was has become known as a performance take off — lighting the burners and going vertical as soon as possible.

One of the highlights from the 2005 show was a group of MiG-21 LanceR aircraft from the Romanian Air Force. The jets were on their way to RAF Lossiemouth to participate in an exercise with the Tornado fleet and stopped at Waddington beforehand. Sadly, the aircraft were not parked in the best location — bang opposite the fun fair!

The ill-fated Nimrod MRA.4 flew at the 2007 airshow. The differences between this aircraft and the MR.2 it was planned to replace were vast — it was so much quieter thanks to its new Rolls-Royce BR700 engines.

2008 — the Vulcan year. Avro Vulcan XH558 returned to the skies in October 2007 and after a few months of test flights and paperwork, public display authorisation was granted just two days before the show in 2007. The aircraft did not fly on the Sunday due to unserviceability issues, but on the Saturday XH558 flew in formation with the BBMF Lancaster PA474 to the delight of the crowds. Seeing and hearing the Vulcan over Waddington was superb.

The eight immaculate and highly polished F-16 Fighting Falcons of the Thunderbirds parked in a very neat line at the end of a successful day's flying. The flags on the sides of the aircraft just behind the cockpit are to show every country the team have displayed in since they were formed in May 1953.

A 25-image montage of just some of the highlights not already covered from 2004-2014.

CHAPTER 16

Preserved Aircraft

Waddington has had a variety of gate guards throughout the past 40 or so years. The main and longest standing one is Avro Vulcan XM607 – the original 'Black Buck' aircraft mentioned previously. It was originally parked opposite the main gate at the end of Mere Road and was only just viewable from the A15.

Eventually, XM607 was moved to a stand at the side of the A15 and it remained here until 2021 when it was towed to be refurbished. After refurbishment, the Vulcan has been parked at various locations, the plan being to keep it mobile and use it for on-base events in the future. There were also rumours that the original stand on the A15 was no longer suitable for it, so we will see what happens there.

XM607 five years prior to its famous raid, performing at the 1977 Jubilee Air Show at RAF Greenham Common. BOB WARD

XM607 in its original spot, taken in August 1987.
GRAHAM ROBSON

XM607 on its stand by the A15 on a summer's evening in June 2020.

XM607 post refurbishment, in September 2023. MIKE KEIGHTLEY

The Victor on gate guard duties, with the Vulcan just out of shot to the left, taken 30th April 1988. BOB WARD

The shortest-lived gate guard was also a 'Black Buck' veteran, in the form of Handley Page Victor K.2 tanker XL189, which appeared for a short time at the end of the 1980s parked behind the Vulcan, rather than in front as it would have been in flight.

During 'Black Buck' One, XL189 was meant to be the second to last tanker, with the plan to pass on as much fuel as possible to the lead Victor that would take XM607 to complete its run, but a thunderstorm during the final refuelling bracket broke the lead Victor's refuelling probe, meaning it was no longer able to receive fuel. Both Victors had already flown further than they intended

XL189 met its sorry fate in 1989. GRAHAM ROBSON

to and now they had to reverse positions with XL189 now being the lead Victor having to receive the fuel from the wounded Victor.

The final top-up of the Vulcan was completed and XL189 turned for home, short on fuel. But it was not until after the Vulcan crew had called 'Superfuse' – the code word for a successful drop, that the crew of the Victor were able to call base back at Ascension for help. Sadly, the aircraft was scrapped, after the powers that be decided bases should only have one gate guard. Nothing large was saved, but it is believed there are some small pieces of it in private collections, somewhere!

23 Squadron received their gate guard some six years after reforming, with Phantom FGR.2 XV497 arriving early morning on 23rd June 2002. The aircraft had been at RAF Coningsby since being the very last RAF Phantom to fly in October 1992. No longer airworthy, the aircraft was brought in by road and lifted by crane over the perimeter fence. It was usually opened during airshow weekends to allow visitors the chance to sit in the cockpit and have photos taken. XV497 eventually left Waddington in 2012 and after some time at former RAF Bentwaters, is now on display at the Norfolk and Suffolk Aviation Museum. IAN HAMPSON

5 Squadron's gate guard came in the form of English Electric Lightning F.6 XR770, in a stunning grey and red scheme. The Lightning was parked outside the squadron headquarters from 2008 until 2015, when the aircraft's owners had to dispose of it. The aircraft now resides at Manston History Museum. GRAHAM ROBSON

8 Squadron had a gate guard in the form of Hawker Hunter F.6A XE606, which was painted in the colours of a previous 8 Squadron Hunter — XE620. The aircraft was unveiled on 2nd June 2013 in the same spot the 23 Squadron Phantom had previously resided. With the arrival of the Red Arrows in 2022 the Hunter was removed and now resides at the North East Land Sea and Air Museum where it will be repainted in the markings of 20 Squadron and ultimately put on display at RAF Boulmer. BENN GEORGE

CHAPTER 17

Heritage Centre

RAF Waddington, like many RAF bases, has a heritage centre detailing the base's history right back to 1916, with many photos, personal accounts, and artifacts. The centre has three rooms — beginning with a briefing room in the centre where you start the tour, which includes many, many photo albums. The second room houses the base's history from the cold war years up until the present day. This room has some items from the Vulcan fleet and an operator's console from the back of a Sentry, plus a few other aircraft parts and covers all types currently operated at Waddington. The third room looks at the early years up until the end of the Second World War, with a rather large, yet incomplete centre piece.

A large metal Lancaster-shape frame has been built to support the fuselage along with a raised platform as there is no forward fuselage from where the cockpit would be. The skin of the aircraft is very thin and despite being in the ground for over 60 years the colours are still very clear, it even being possible to see the brush strokes of the red squadron lettering — PD259 was coded JO-G.

Avro Lancaster PD259 was operated by 463 Squadron Royal Australian Air Force and sadly crashed on 31st August 1944 in Kingussie, Inverness, Scotland, without any survivors. The aircraft came down in a peat bog after breaking up at around 10,000ft while on a training mission; the cause of the break-up is believed to be the g-force on the aircraft after the pilot tried to pull out of a dive.

The aircraft was relatively new when it crashed, having flown fewer than 60 hours. As much of the aircraft as possible was recovered in June 2009 and due to the remote location of the crash site, a Griffin HAR.2 helicopter was used to winch up parts and load them on to a nearby lorry.

Although not part of the Heritage Centre, Waddington also has a Memorial Garden which can be accessed with the help of Heritage Centre staff. The garden has multiple memorials, the centre of which is a propeller from the Lancaster that is in the centre. Beside the propeller is a plaque dedicated to the memory of all those who died while flying with 463 and 467 Squadrons.

There is also a memorial for 44 Squadron, one of Waddington's longest residents and one for 31 Squadron, one of Waddington's newest residents, with the memorial moving from RAF Marham. There are a few small plaques as well, arranged either side of the Lancaster propeller. These were unveiled on 12th October 2021 during a station memorial day for those who had sadly passed during the previous 18 months, where a formal service could not occur due to Covid 19.

Families of those who passed away were invited onto the base and met with the station commander in the officers' mess. They were then taken to the Memorial Garden by cars driven by the station's MT section along High Dyke. Military personnel in their Number One uniforms lined the road for about half a mile, it was a very impressive sight. I was present to remember my father-in-law — Warrant Officer Steve Hill, who had passed away in December 2020. He had left the RAF in April 2019 after 40 years' service and joined the Royal Auxiliary Air Force.

CHAPTER 18

Lincolnshire and Nottinghamshire Air Ambulance Charitable Trust

The Lincolnshire Air Ambulance Trust was formed in 1993, with the Air Ambulance becoming operational in April 1994 and being based at RAF Waddington. The trust started out with an ex-Police Air Service MBB B105 helicopter, registration G-PASC and operated from what was known as Bravo dispersal, but had limited time airborne due to lack of funds, often being grounded for months at a time.

In 1995, thanks to sponsorship, the helicopter was able to operate 365 days a year and the following year the service was expanded to cover Nottinghamshire as well, becoming the Lincolnshire and Nottinghamshire Air Ambulance. The MBB B105 served until 2000 when it was replaced with an MD902 Explorer G-LNAA which in turn was replaced 10 years later by another MD902, G-LNCT.

The trust said goodbye to the MD902 in 2017, with an upgrade to an Agusta Westland AW169, G-LNAC which was a much bigger helicopter – allowing the paramedics on board better access to patients. The current helicopter, AW169 G-LNCC arrived in July 2021 and is a top-spec piece of kit, able to fly at night and in almost all weathers. Throughout the ten years of Waddington Airshows, the current Air Ambulance would usually be on static display next to the charity's stand to try bring in more income. Currently it costs £10m to run the Air Ambulance each year.

In May 2021 the Air Ambulance trust opened their own purpose-built headquarters and base for the helicopter, along with a

The current LNAA Helicopter — AW169 G-LNCC on static display at RAF Waddington's families day, 2023. MATT HALLAM

Helimed 29 on finals to land at Kookaburra, with the magnificent new building in the foreground, taken 28th February 2024.

helicopter landing pad, which enabled all elements of the charity, including the critical care cars to operate under one roof. The trust was lucky to receive a grant which enabled them to afford the building — the money coming from the cancellation of the Northern part of the HS2 rail network.

The building is located on one of the original 1940s bomber dispersals, on the opposite side of the A15 to RAF Waddington. The helicopter pad has been given the name 'Kookaburra' with pilots asking to land and depart from this unusual name when flying. The reason for the name goes back to the 1940s again. 467 Squadron were a Royal Australian Air Force Lancaster squadron, who operated from the dispersals on the east of the runway and A15.

One of the squadron's Lancasters was coded 'K' Kitty and was known as Kookaburra. The dispersal where Kookaburra used to park is exactly where the yellow 'H' on the landing pad is now, over 80 years later. It was hoped that the Air Ambulance could have some sort of celebration in 2024, being its 30th anniversary.

The author next to one of the Hawk T.1As of the Royal Air Force Aerobatic team — the Red Arrows, parked on the bays at RAF Waddington towards the end of research on this book.